John George Wood, George B. Sowerby

The Common Objects of the Sea Shore

including hints for an aquarium

John George Wood, George B. Sowerby

The Common Objects of the Sea Shore
including hints for an aquarium

ISBN/EAN: 9783337271985

Printed in Europe, USA, Canada, Australia, Japan

Cover: Foto ©Andreas Hilbeck / pixelio.de

More available books at **www.hansebooks.com**

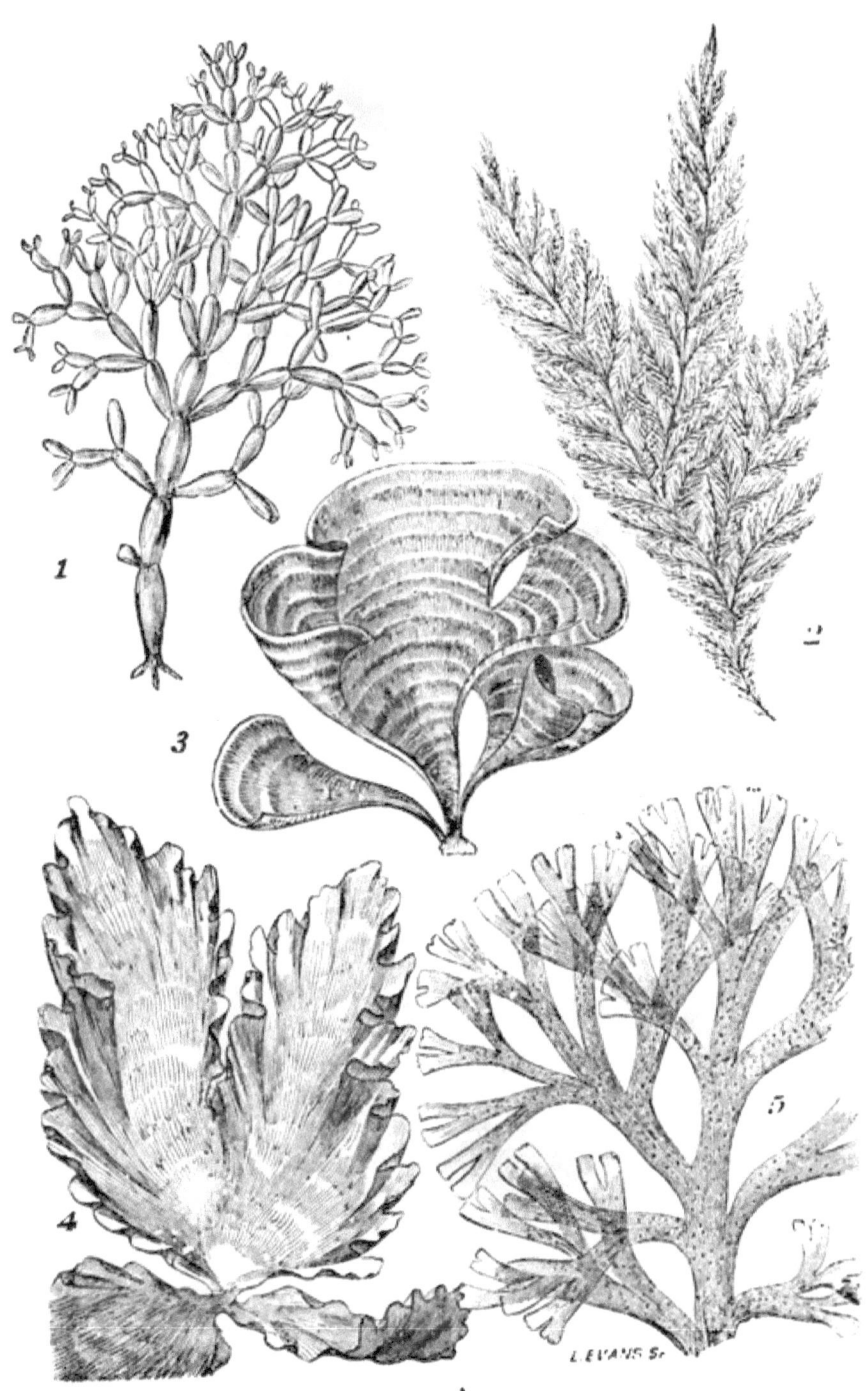

A

THE COMMON OBJECTS

OF

THE SEA SHORE;

INCLUDING

HINTS FOR AN AQUARIUM.

BY THE
REV. J. G. WOOD,
AUTHOR OF "THE ILLUSTRATED NATURAL HISTORY."

WITH ILLUSTRATIONS BY G. B. SOWERBY.

LONDON:
ROUTLEDGE, WARNE, & ROUTLEDGE,
FARRINGDON STREET.
1860.

NOTICE.

An Edition of this Work, printed in large type, with the Illustrations Coloured, bound in cloth, may also be obtained.

LONDON.
R. CLAY, PRINTER, BREAD STREET HILL.

PREFACE.

THIS little work is simply a popular account of the "Common Objects of the Sea-Shore," and is restricted to those objects which every visitor to the sea-side is sure to find on every coast. For descriptions of those creatures which only inhabit certain localities, and those whose lives are passed in the deep water, requiring the dredge, the net, or the drag to bring them to the light of day, the reader is referred to those magnificent and comprehensive works that have been written for tne purpose of illustrating particular branches of science.

During my visits to the sea-coasts for the last six or seven years, I have taken note of the questions put to me by persons who were anxious to know something of the curious objects that everywhere met their eyes; and the following pages are, as nearly as possible, the condensed conversations that then took place.

The whole of the illustrative plates were drawn expressly for this work by Mr. Sowerby, whose name is a sufficient guarantee for their truth.

LONDON, *May*, 1857.

COMMON OBJECTS OF THE SEA-SHORE.

CHAPTER I.

MARINE BIRDS—PORPESSE.

Whether the sea is approached by land or by water, the first indications of its existence are generally to be found in the air. On some days, the electric clouds that skirt the cliffs map out, as it were, the sea-coast; and when such signs fail, the marine birds give evident tokens that the sea, their great store-house, is close at hand. With the birds, then, we will commence our observations of the sea and its shores.

The bird that usually presents itself as the ocean's herald is the Common Gull (*Larus canus*). There are some twelve or thirteen species of British Gulls, including the Kittiwake and the Iceland Gull. The bird represented in the accompanying figure is the Great Black-backed Gull (*Larus marinus*), which is tolerably frequent on our coasts, but not so often seen as the Common Gull, nor does it form such large societies as those in which its more sociable relations love to congregate.

Why the word "gull" should be employed to express stupidity I cannot at all comprehend, for the gulls are very knowing birds indeed, and difficult to be deceived. If a piece of bread or biscuit be thrown from a boat, it remains but a very short time on the surface of the water before it is carried off by a gull, although previously not a bird was visible. But if a number of gulls are flying about, and a piece of paper or white wood be thrown into the water, there is not a gull who will even stoop towards it, although to the human eye the bread and the paper appear identical. The cry of the gull is very curious, being a kind of mixture of a wail, or scream, and a laugh, and on a dark stormy day adds wonderfully to the spirit of the scene. Its flight is peculiarly quiet, combining great power of wing with easy elegance of motion. It is a very bold bird, and for many miles will follow boats so closely that the very sparkle of its eyes is plainly visible, as it twists its wise-looking head from side to side while watching the voyagers.

BLACK-BACKED GULL.

The gull is an exclusively marine bird, being found only on the sea-shore or at the mouths of large rivers, although more than usually violent storms occasionally drive it inland, where it wanders about for some time very miserable, and quite out of its element, until it gets shot by some rustic sportsman.

The next bird to which our attention will be directed is the Common Tern (*Sterna hirundo*), or Sea Swallow,

TERN.

as it is very appropriately called. It belongs to the gull family, and has many of the gull habits, but is readily distinguishable from the gulls, even at a considerable distance, on account of its rapid, darting flight. It is not at all unlike a swallow in general shape, for the wings are long and pointed, the body is rather large in front, and tapers to a point, and its tail is forked like that of a swallow.

It is extremely dexterous in its capacity as a fisher, for in its swift flight over the waves it darts upon any small fish that may be unfortunate and curious enough to come near the surface, and scoops it up, as it were, from the water, without seemingly interrupting the speed of its course. The nest, if it can be dignified with the title, is merely a hollow scooped in the sand, well above high-water mark; and in this hollow two or three eggs are deposited. On the Scottish and northern coasts the Common Tern is not often found, but the

Arctic Tern (*Sterna macroura*) comes to supply its place. There are ten or eleven species of British Terns.

On many of the English coasts, especially those that look towards the Channel, may be seen a tolerably large black-feathered bird, having a yellowish countenance, a decidedly long and rather hooked bill, and a pair of green eyes. This is the Cormorant (*Graculus carbo*)

one of our three British representatives of the Pelican family. The enormous pouch which decorates the lower bill of the white pelican is only rudimentary in our British pelicans, probably because there would be no use for it, as the birds live on or close to the coast.

The other English pelicans are the Gannet, a figure of which will be given shortly, and the Common Shag, a bird of a monosyllabic English cognomen, but who ought to consider himself recompensed by the scientific name given to him by certain naturalists, namely,

Phalacrocorax cristatus; the epithet *cristatus*, or crested, being due to a tuft of reverted green feathers that decorates the head. This tuft, however, is only worn during the breeding season, when most animals put on their gayest apparel, and is lost as soon as the young *Phalacrocoraces cristati* take their places as independent members of society.

The cormorant is a persevering fisher, insatiable in appetite, and almost unparalleled in digestion. The pike and the shark among fish appear to possess much the same proportionate digestive power as the cormorant among birds. The cormorant is not content with sitting, like the heron, on the edge of the water, and snapping up the fish that may enter the shallows; or even, like the gulls, with seizing them from the surface of the waves; but he boldly defies his prey in its own element, plunges into the water, dives below the surface, and actually proves himself a more expert swimmer than the very fish themselves. In former days the cormorant was employed in England for the purpose of catching fish; and such is still the case in China. The Chinese cormorant, however, is not the same species as that which is found on our coasts. It is rather a curious circumstance, that one of the mammalia, namely, the otter, and some of the birds, should be enabled to carry on a successful subaqueous chase, and that both beast and bird have been pressed into the service of man.

The cormorant is sometimes found inland, especially in the winter season, and exhibits its powers among the fresh-water fish.

Although the pouch is comparatively small in the cormorant, it still exists, and is useful in giving elasticity to the throat and neck; a property which is much required, for the cormorant is a very greedy bird, and often swallows fish of so large a size that a throat of twice its dimensions seems incapable of permitting the passage of so bulky a body. In order to swallow a fish, the cormorant generally seizes it crosswise, tosses it in the air, and then catches it as it descends with its head downwards. One of these birds, however, has been seen to miss its aim, and to catch the fish with its head upwards; in this position the cormorant endeavoured to swallow its prey, but when the fish had passed about half-way down its captor's throat, the sharp fins prevented its further progress, and both bird and fish were soon dead. The poor cormorant seemed terribly distressed, and made violent struggles, but all to no purpose; for the fish was immovably fixed, and could neither be swallowed nor rejected.

The feathers of the cormorant, although they appear to be of a dusky black, are really of a very deep green, so deep, indeed, as to appear black at a little distance, something like the plumage of the magpie. The nest is composed of dried sea-weed, and is usually placed on lofty rocks, but is sometimes built among the branches of trees. The eggs are remarkable for a thick coating of chalk, which seems to envelope the shell quite independently, and can be easily removed with a knife. There are from three to five eggs in each nest.

THE GANNET.

Our remaining English pelican is the **Gannet** (*Sula bassanea*), also called the Solan Goose. It is to be found on many of our shores, but especially on the well-known Bass Rock, at the entrance of the Frith of Forth. There is no difficulty in identifying this bird, even at some distance, as it has very much the appearance of wearing spectacles; a circumstance that has earned it the title of the Spectacled Goose, although it is not a goose at all.

It is in the search after these birds, their eggs and young, that the St. Kilda cragsmen imperil their lives year by year.

The Auk family find representatives in the Guillemot, and the comical little Puffin. The Guillemot (*Uria troile*) is a common bird on many of our coasts, and

GUILLEMOT.

may be seen, in the breeding season, sitting with extraordinary gravity and importance over a solitary egg.

The egg is often laid, and the young hatched, on such a narrow ledge of rock, that it is quite a wonder how the egg can escape a fall, or how the young bird can even open its big beak without toppling over the precipice.

The guillemot has earned the epithet of "foolish," because, when sitting on this solitary egg, it will suffer itself to be taken by hand, rather than forsake its duty. I would suggest that the word "faithful" be substituted for "foolish." The egg is a very handsome one, very large, and variable both in colour and shape. It is generally covered with large irregular blotches, of a brown colour, on a pale-green ground.

As to the Puffin (*Fratercula arctica*), it is generally to be found in company with the guillemots, and indeed lives in much the same manner. It is a lively little bird, easily distinguished by its large beak; from which feature it has derived the popular name of Sea Parrot. This beak is very useful to the bird for three especial purposes: the first being, to catch fish; the second, to dig the burrows in which its egg is laid; and the third, to fight the ravens and other foes who try to get at the egg.

PUFFIN.

The favourite food of the puffin appears to be the common sprat, which it chases under water, and of which it generally secures six or seven, all arranged in a neat row along the puffin's beak, and hanging by their heads.

There are many more marine birds that are often seen, but those already mentioned are the most common. Yet there is one other bird that I must notice, because it has not so much the marine aspect as the gulls and cormorants: this is the Dunlin Sandpiper (*Tringa*

DUNLIN.

alpina), a very interesting little bird, that frequents the sandy shores in great numbers, for the purpose of feeding on the insects and small crustaceans that are found in such profusion, either buried in the sand, or hidden under stones and drifted sea-weeds.

It is quite aware that on the edge of every wave may be found the various substances which constitute its food, and so skirts the very margin of the sea, running hither and thither; and occasionally venturing a few paces into the retiring waters in chase of some detached limpet, some houseless worm, or tiny crab, as restlessly, and almost as untiringly, as the many-voiced waves themselves.

There is no difficulty in watching the habits of this, or indeed of any other bird. All that is required is perfect stillness and silence, and the birds will come and pick up their food almost within arm's reach.

In the hot summer months the observer may watch the sands without seeing a single Dunlin, for they then desert the sea-shores in favour of inland moors, where they lay their eggs, and hatch the young; returning with their offspring towards the end of August. This bird is sometimes called the Purre.

If we now leave the sands for a time, and ascend the cliffs, we shall probably be indulged with a transient glimpse of a very singular animal. Some little distance from the shore a number of black objects may be seen partly emerging from the water, executing a summersault, and disappearing below the surface. These

PORPESSE.

are Porpesses (*Phocæna communis*), and very curious creatures they are, belonging to the mammalia; forced, therefore, to breathe atmospheric air, and yet permanently inhabiting the sea, with something of the form and many of the habits of the fish. It is a curious connexion of the two most distant links of the chain of the Vertebrates, the mammalia being the highest, and the fish the lowest.

Some people say, that as it looks like a fish, and lives like a fish, to all intents and purposes it *is* a fish. So it is, if the diver at the Polytechnic Institution is a fish; for it holds its lease of life on precisely the same tenure. Both diver and porpesse must breathe atmospheric air, or they would die; and therefore each finds means to supply himself with that indispensable material. The diver surrounds himself with a supply of fresh air, with which to renovate his blood; but the porpesse is able to renovate a surplus amount of blood, that lasts him for some time : so the chief difference is, that the diver takes down with him oxygen externally, and the porpesse internally. The man goes down inside the diving-bell, but the diving-bell goes down inside the porpesse.

Yet the porpesse has no reservoir in which atmospheric air is retained, for such a formation would make it too buoyant. There is, however, in the *cetaceæ*, to which family the porpesse belongs, a reservoir of blood, which is renovated by the atmospheric air, and is passed into the system as required. Even man has the same power, although in a limited degree. In general, a man cannot hold his breath more than one minute, and it is not every man who can do even that. But if he thoroughly renovates his blood by expelling all the impure air that remains in the minuter tubes of his lungs, and takes a succession of deep inspirations, he will be able to abstain from breathing for a much longer period. I have just made the experiment myself, and held my breath without difficulty for a minute and a-half ; and had there been any necessity, could have done so for another half minute.

The porpesse is rather a sociable animal, being generally seen in shoals, or schools, as the sailors call them. I should hardly have said so much about so common a creature, were it not for the purpose of pointing out these remarkable facts in its structure and habits : and even though it be common, it is not so well known as might be imagined. Not long ago, as I was on board a steamer, a worthy old lady began to exhibit symptoms of nervousness and alarm. I thought she was fearing a storm, and told her that there was not the least danger of any commotion of the elements, for the barometer had been steadily rising for the last two or three days. However, it was a different subject that caused her uneasiness. She had heard that there were porpesses in those parts, and wished to know whether we were likely to meet one. I told her that we probably should meet several, but not so many as if a gale were impending. At this reply her fright evidently increased, and she asked, in much trepidation, whether, if we did meet one, it would upset the ship!

CHAPTER II.

WHELK—COWRY—COCKLE—PHOLAS—LIMPET—SEA-WEEDS ON SHELL—BALANUS—PURPURA—MUSSEL—PERIWINKLE, YELLOW—PERIWINKLE, COMMON—TROCHUS.

DESCENDING to the shore, we shall probably see at our feet many shells, or fragments of shells, which have been washed upon the beach by the advancing tide, and which having lodged behind a stone, or being sunk into the wet sand, remain behind when the waves retreat. These shells are almost invariably empty, their inhabitants having either died a natural death, or having fallen victims to some ravenous inhabitant of the sea.

The strong house with which most of these creatures are furnished would seem to be an effectual defence against the efforts of open foes, while the sensitive nervous nature with which they are gifted would appear to secure them from insidious attacks. Yet the hard, stony shells, that turn the edge of a steel knife, are constantly found to be perforated by creatures that can be squeezed flat between the fingers, and whose bodies are no harder than the human tongue. Formerly, the external characters of shells were the only object of the collector; and the conchologist, as he was termed, might have, and very often did have, a large collection of valuable shells, without the least idea of the form, food, habits, or development of the creature that secreted them. Now, however, those who examine a shell are not satisfied unless they know something of the creature that inhabited it, and from whose substance it was formed: and so this branch of Natural

History has leaped at once out of the mere childish toy of conchology into the maturer science of malacology. The former treated merely of shells, and therefore excluded the vast army of molluscs, that wear no shells at all; but the latter treats especially of the animal, considering the shell to be of secondary importance.

And yet, even though the shell is considered to be inferior to the animal by whom it was secreted, much more attention is paid to the shell itself than was the case in the old conchological times. In those days the mere shape and colours of the shell were the characteristics by which its name and place in the system were determined; but now we submit the shell to the searching powers of the microscope, and find that various kinds of shells are characterised by various arrangements of particles, and are acted upon by polarized light in various ways. It is, therefore, quite possible to fix the character of a shell from a single fragment no larger than a pin's head. There are few things more curious than this wonderful arrangement of the particles; which, by the way, are brought within the scope of the microscope, by making very thin sections of the shells, by the aid of saw, file, and hone.

One of the commonest shells found on the sea-shore is the Limpet (*Patella vulgata*). See plate B, fig. 3.

In its living state it may be found adhering closely to rocks or other substances, that give it a firm basis of support. The adhesion is caused by atmospheric pressure, for the limpet is enabled to raise the centre of that part of the body that rests on the rock, while the edge is closely pressed upon it. This movement causes a vacuum; and so firmly does the air hold the limpet in its place, that the unaided fingers will find great difficulty even in stirring it. The firmness of adhesion is also increased by the fact that after the animal has remained for some time in one spot, it forms a hollow in the place where it rests, corresponding in size with the shape of the shell. Into this depression the shell sinks, and consequently there is no possibility of reaching its edge, where alone it is vulnerable. When, however, it is not warned, and prepared for resistance, it can be easily detached by a sharp movement of the hand.

In general, it is not a migratory creature, and, consequently, is often seen to be so covered with parasites of various kinds, that its form can hardly be recognised. I have now in my aquarium a limpet-shell, on which a specimen of the common laver (*Ulva latissima*) and another of *Porphyra laciniata* have affixed themselves, and are growing luxuriantly. There was also in the same tank another limpet-shell, on which was growing a whole forest of sea-grass (*Enteromorpha compressa*), expanding as widely as the crown of a man's hat. The acorn barnacle, too, often takes possession of the limpets, and it frequently happens that, in some dark cavity of rock, a colony of limpets may be found, each so covered with these sessile barnacles, that not a particle of the original shell is visible. Of this, however, we will speak hereafter.

The figure, plate B, fig. 3 *a*, represents the appearance of the limpet as it is generally seen on the rocks; 3 *b* represents the under-surface of the same object, and shows the animal itself. The limpet may easily be thus seen, if it is placed in a vessel of sea-water with flat glass sides, for it soon crawls up the side, and so exhibits itself very perfectly.

The Common Whelk (*Buccinum undatum*) is another shell that is sure to be found on the sands. This is so well known a shell that no particular description is here necessary, but mention will be made of it on a succeeding page. The little shell, figured on the left side of the whelk, is one of the cowries, of which there are almost innumerable varieties. Some of them, found in the tropical seas, are of very large size, while others are much smaller than the specimen represented. One species of this shell is used as a medium of exchange in some countries. Money that can be picked up on the sea-shore is, however, of very small value, fifteen hundred cowries being considered as an equivalent to

COWRY. WHELK.

one English shilling—hardly reimbursing the collector for the trouble of stooping so often.

There is another shell allied or distinctly related to the whelk which is very common on our coasts, and which is well worthy of notice. This is the *Purpura lapillus* (plate D, fig. 4), a shell that is sometimes found nearly white, but mostly banded with brownish orange, as is represented in the figure. Now, the creature that inhabits this shell is one of those animals that furnished the famous purple of the ancients, and from that property it derives its name of Purpura. The colour is not particularly beautiful, and it is rather remarkable that the ancients, who had very good taste in colours, should have placed so high a value on this purple, which, according to their own account and our observations, closely resembled clotted blood.

Perhaps, however, its rarity constituted its value; for there is so little in each shell, that an enormous number of victims must have perished before a sufficiency of the dye for one robe could have been obtained.

The ancients seemed to have managed the extracting process in rather a clumsy manner; but it is easy enough to procure the dye without mixing it with the juices of the animal, as seems to have been the case in the olden times. If the reader would like to try the experiment, it may be done as follows:—

Let him look among the rocks at low water, and plenty of the shells may be found tolerably close together. When a sufficient number are collected, they should be killed by placing them in fresh water, after the shell has been pierced or broken, as otherwise the animal shuts itself up so tightly that the water cannot gain admittance. When the creatures are quite dead, the colouring matter may be found in a yellowish-looking vessel, that derives its colour from the substance contained within. There is very little of this colouring matter in the vessel. Now, if this yellow substance be spread on white paper and placed in the sunshine, a blue tinge enters the yellow, making it green. The blue gradually conquers the yellow, and the green soon becomes blue. Another colour, red, now makes its appearance in the blue, and turns it into purple. The red becomes gradually stronger, and in its turn almost vanquishes the blue, but does not quite succeed in doing so; for the blue, having taken so much pains to turn out the yellow, will not entirely vacate the premises, and, coalescing with the red, forms a deep purple, the red very much predominating. So we have here all the primary colours fighting for the dominion, and yellow, the most powerful of the three, forced to retire before its complementaries.

There are great numbers of little shells, called Tops from their form, which are found plentifully on every coast, either empty and cast ashore by the waves, or living, and found adhering to the sea-weeds that are laid bare at low water. It is not often that these shells are found quite perfect, for the shell is generally worn away at the apex, so that the colouring substance is removed and the point of the shell is white. One of the most beautiful of these shells, the Livid Top (*Trochus ziziphinus*), is represented on plate B, fig. 1.

The tongue of this species is remarkable for its

structure. Many molluscs are furnished with **very** wonderful tongues, the true beauty of which can only be seen by placing them under a microscope of moderate power. Their tongue is easily extracted by drawing it out from its hiding-place with a needle, and cutting it off—the owner being, of course, previously killed.

When this organ is properly displayed, it will appear furnished with one array of teeth, very minute, but very strong, and quite adequate to the work which they have to perform. In fact, the tongue is a miniature file, and is used not so much for tasting the food, as for a rasp, wherewith to cut it off. The top, therefore, is an useful inhabitant of an aquarium, for he saves an immensity of trouble in keeping the glass sides clean. After an aquarium has fairly settled itself, the algæ pour out their spores, and these, adhering to the glass, there affix themselves, so that in a few weeks the glass becomes dimmed by the mass of minute vegetation. Here the tops and periwinkles come to our aid, and by means of the natural scythes with which they are armed, soon mow away the greater part of this vegetable growth. They seem to do their work as composedly and regularly as if they were paid by the day for it. The Livid Top may be found alive among the rocks at low water.

I have already stated that the periwinkles are useful inhabitants of an aquarium, and such is the case as long as they can be kept alive. But they are often very perverse in disposition, and show greater predilection for dying than for mowing. The Common Periwinkle, so familiar in our streets, is tolerably hardy in confinement, and may be kept for some time. But the handsomer Yellow Periwinkle (*Littorina littoralis*), which is represented on plate B, fig. 2, is still more delicate in constitution, and seldom survives for many weeks. But even the Common Periwinkle is a pretty creature, as it exhibits itself when crawling upon the glass of the aquarium, or on the sea-weeds where it finds its food. The body is prettily banded with multitudes of narrow dark markings, and the mode in which the creature slides itself over the glass is very curious.

There is a very pretty shell found in tolerable profusion on our sands, and which will be recognised at once from its portrait (plate B, fig. 5). This is the Common Wentletrap (*Scalaria communis*). It is not only a pretty shell, but holds relationship with a very aristocratic connexion. The Wentletraps are divided into two great sections; the false Wentletraps, the whorls of whose spires touch each other; and the true Wentletraps, whose whorls are disjointed from each other. Of the former section our little friend is a good example; and of the latter, the aristocratic relative alluded to. This is the Royal Staircase Wentletrap, a shell formerly of such rarity that a specimen only two inches and a quarter in height would fetch eighty or ninety pounds.

The next shell which I shall mention is the Common Cockle (*Cardium edule*), represented on plate B, fig. 6.

Perhaps this is the most abundant of all the littorine shells; for if a handful of shells be gathered at random from the sands, nearly one-third will be cockles. When living, the animals find a home under the sand, in which they lie buried. The cockle is a capital delver, and, armed with his natural spade, digs for himself a hole in the sand nearly as fast as a man can dig with a spade of metal. As for the wooden spades, so much in vogue on sandy coasts, they have hardly a chance against the cockle.

Many an observer has been perplexed at the little jets of mingled sand and water which are so often seen issuing from the sand when the waves have retired. These tiny geysirs are occasioned by the cockles that lie buried beneath the sand, and which are still in the water below the sand level, although the surface is tolerably dry.

Our cockle, however, is not only a digger, but a

jumper, and the same instrument which serves him as a spade to dig a hole in the sand also serves him as a foot by means of which to spring into the air

There is another burrowing shell, that is found on most sandy beaches. This is the Razor-Shell (*Solen ensis*), for a representation of which, see plate B, fig. 7.

This creature burrows even deeper than the cockle, being often found at the depth of two feet. It does not, however, seem fond of sinking thus low, but generally remains sufficiently near the surface to permit the tube just to project from the sand. The burrow in which the animal lives is nearly perpendicular, and in it the Solen passes its entire life, sometimes ascending to the surface, and sometimes descending to the bottom of its burrow, for it has none of the locomotive faculties of its fellow-miner, the cockle. But although its range of travel is circumscribed, the narrowness of its habitation is compensated by the activity of its movements therein. The fisherman who wishes to capture the creature is aware of its agility, and takes measures accordingly. As the tide retreats he watches for the jet of sand and water which the animal throws into the air when alarmed by its hunter's footstep. Into the hole from which the jet ascended the fisherman plunges a slender iron rod, which having a barbed, harpoon-like head, pierces the animal, and retains it while it is dragged from its hole. If, however, the fisherman takes a bad aim, and misses his cast, he does not try a second with the same creature, knowing that it will have retreated to the termination of its burrow, whence it cannot be extracted.

Yet another burrowing shell. In most chalky rocks, such as those of which the white cliffs of old England are composed, many portions run well out to sea. If these are examined at low water they will be found to be perforated with numerous holes, running to some depth, and varying considerably in dimensions. These holes are made by the *Pholas dactylus*, plate B, fig. 9, one of the most remarkable animals in creaturedom.

Hard rocks and timber are constantly found perforated by this curious shell, but how the operation is performed no one knows. It is the more wonderful, because the shell is by no means hard, and cannot act as a file. Indeed, in some species, the external shell is almost smooth. And, moreover, if the shell were used as the boring-tool, the hole would be nearly circular, instead of being accommodated to the shape of the shell, as is seen to be the case. However they get into the stone, there they may be everywhere found, and it does not seem to be of much importance whether their habitation be limestone, sandstone, chalk, or oak. Even the Plymouth breakwater, solid stone as it is, was very soon attacked by these creatures.

They are especially obnoxious to the builders of wooden piers, for they seize on the submerged portion of the piles on which the pier rests and do their utmost to reduce them to a honeycombed state with the least possible delay. Lately, however, the Pholades have been conquered; for they cannot pierce iron, and it is found that if iron nails are closely driven into the submerged portion of wooden piles, they bid defiance to the Pholas.

The specimen represented in our figure is shown resting in its rocky bed, and seen edgeways. At each side may be seen the furrowed shells; the foot appears in the centre, surrounded by the mantle, and the tube is seen projecting far beyond the shell. Very many good specimens may be obtained by splitting open the piece of rock, and thus the shells extracted without injury from the rocky home where they have lived and died. In the interior of a perfect shell may be seen a very curious projection, formed something like a spoon. Its object does not seem to be very clearly ascertained. The tube, which has been so often mentioned, is generally a composite organ, composed of two tubes or siphons, as they are called, which are placed closely together, something on the principle of a double-barrelled gun, or an elephant's trunk. Through these tubes passes the water which is necessary for respiration, being received into one tube, drawn from thence over the gills, and finally expelled from the other tube.

There is another boring mollusc, which is on many accounts worthy of notice. This is the so-called Ship-worm (*Teredo navalis*), a representation of which may be found on plate F, fig. 3. It has been placed on the same page with some of the worms, in order to show its very great external resemblance to some animals of that class, and especially its similitude to the Serpula. So closely, indeed, does it resemble the last-named creature, that even Linnæus placed the Teredo between Serpula and Sabella in his "System of Nature."

But this is really one of the molluscs, and a very curious one. It is called the Ship-worm because it has so powerful an appetite for submerged wood, and especially for ship-timber. I have now by me a large piece of oak, the remains of some wreck, which I found entangled among the rocks at low water. It is so completely devoured by the Teredo, that it is almost impossible to find any portion of the wood that is thicker than the sheet of paper upon which this account is printed. Timber, however, can be protected from the Teredo by a closely studded surface of broad-headed iron nails. These nails soon rust through the action of the salt water, and the whole of the timber is rapidly covered with a thick coating of iron rust, a substance to which the Teredo seems to have a strong objection.

The *Teredo navalis* is not a very large animal, but it has a huge overgrown relation, the Giant Teredo, whose diameter at the thickest part is three inches, and its length nearly six feet.

On plate B, fig. 8, may be seen a shell, which will probably be recognised at once as the Common Mussel

(*Mytilus edulis*). The specimen figured is a young one, and is shown as it appears when adhering to the rock by means of the natural cable—or byssus, as it is scientifically named—with which these creatures are furnished.

These shells are exceedingly common, and large masses of them may be found clinging to any rocks or stones where they can anchor themselves. This mussel is called Edulis, or eatable, because it is largely used as an article of food. But it is by no means a safe edible, as at certain times, or to certain constitutions, it acts as a poison, producing most alarming and sometimes fatal effects.

The byssus is an assemblage of delicate, silky, and excessively strong fibres, the origin of which seems to be at present rather obscure. Many shells are furnished with this substance, which is shown in perfection in the great Mediterranean Pinna, some specimens of which measure nearly two feet in length. The byssus of these creatures is often spun and woven like silk, and in many places may be seen gloves, purses, and other objects, which have been made from this substance. It is, however, too rare to be put to any practical use.

The Common Scallop (*Pecten Jacobæa*), generally known in connexion with oysters, may be found abundantly on our shores. Even the empty shells are pretty enough to attract observation; but the animals are more beautiful than their shelly habitation. A living scallop is well worthy of notice, if it were only for the row of eye-like points which are seen

SCALLOP.

peeping out from the very margin of the shell, when the creature holds the **valves partially open**. Whether these brilliant spots are really eyes or not has not been clearly ascertained, but at all events there appears no reason why they should not be eyes; and so to us eyes they shall be.

The scallop is capable of changing its position, and does so by the forcible ejection of water from a given point. This mode of progress is analogous to that employed by the larva of the dragon-fly. The title Jacobæa is given to the scallop, from the shrine of St. James (Lat. *Jacobus*), at Compostella; to which spot journeys were made by pilgrims, who, in token of having paid their devotions at St. James's shrine, wore a scallop-shell in their hats for the admiration of their contemporaries, and bore it on their coats-of-arms for the information of their posterity.

The story which connects the scallop-shell with St. James is very curious, but too long for insertion.

The last shell-bearing mollusc which I shall mention is one which does not at first appear to be a mollusc at all. This is the curious little Chiton, a creature which, instead of a tubular shell like the Teredo, a single whorled shell like the whelk, or a double shell like the scallop, bears an array of eight shelly plates on his back, and thus gives to the observer an idea of a tiny marine armadillo.

The entire back of the Chiton is covered with a strong leathery coat, much larger than the living centre of the animal. Upon this leathery mantle are placed eight shell-plates, which overlap each other just as do the tiles of a house. They are not very large on our English coasts, but some foreign species are found which exceed four inches in length.

If the shell-bearing molluscs are remarkable for the elegant form and brilliant colouring of their habitations, they seem to be equalled, if not eclipsed, in beauty by a race of molluscs which possess no shell at all, and whose chief beauty is derived from the singular peculiarity of formation from which their name is derived. These **are** the Nudibranchs, or Naked-gilled Molluscs, so named because their respiratory apparatus, instead of being concealed within their bodies, or defended by shells, is placed upon the exterior, in apparently heedless defiance of surrounding objects. And the more that the delicate construction of these branchiæ is seen, the more wonderful does it appear that these organs should be placed in the position which they occupy without suffering serious injury. If the lungs of one of the mammalia were to be attached to its sides, and permitted to hang loosely therefrom, exposed to the invasions and collisions to which they would probably be liable, the owner of the said lungs would hardly feel comfortable. But the lungs, gills, or branchiæ of the mollusc are so exceedingly delicate, that the mammalian lung appears quite coarse by their side.

There are many species of Nudibranchs found on our coasts, one of the commonest of which (*Doris ptilosa*) is represented on plate N, fig. 4. The gills may be seen spreading like a feathery plume, or a radiating flower, on the upper surface of the creature. The position of the branchiæ is by no means uniform, for indeed the most fertile imagination would hardly venture to depict such fantastic forms as are found among the Nudibranchs, or, if they were depicted, could hope that such wondrous shapes should be received by men as existing in the same world with themselves.

Some species, like those whose shape has been already alluded to, are nearly flat, and wear their lungs much as a gentleman wears a bouquet, in his button-hole. Others have their lungs neatly arranged round their bodies in little spreading tufts, so that the creature has something the aspect of a floriated coronet. Some have their whole dorsal surface thickly studded with lungs, so that it would bear a decided resemblance to a hedgehog, were it not that the spikes must be semitransparent, and tinged with the most exquisite colours. Again, there are some species which carry their lungs at a distance from their bodies, and present them to the waves as if they were holding the branchiæ in the hands of their outstretched arms; while there are some whose forms are so utterly unique and grotesque, that a description would be useless except it were accompanied by a drawing.

As to the colouring of these creatures, there is hardly a tint, from blackish grey to the most brilliant carmine, that is not found in some member of this strange family. They all belong to that division of the molluscs that go by the name of Gasteropoda, because the lower surface of the body forms the foot by which they move from place to place. By the aid of this foot they often float on the surface of the water, as has been already recorded of other molluscs. This action, however, has been well described, as creeping on the superincumbent stratum of air. Many species of the genera Doris and Eolis, together with others, may be found, at low water, clinging to the rocks and stones. They will hardly be recognised as Nudibranchs at a hasty glance, for they subside into shapeless gelatinous knobs as soon as the waves leave them, and do not resume their expanded form until the surging sea returns.

The Nudibranchs, although most lovely creatures, are very unsafe inhabitants of an aquarium, in spite of their delicate and dainty looks; and a wolf would be about as appropriate an inmate of a sheepfold, as a Nudibranch of an aquarium where sea-anemones live. Even the giant crassicornis, or Thick-horned Anemone, has fallen a victim to the insatiable appetite of these greedy creatures. In closing this short description of the Nudibranch, let me strongly recommend the reader to examine, if possible, the beautiful work on these creatures by Messrs. Alder and Hancock, published by the Ray Society.

CHAPTER III.

MARINE ALGÆ, OR SEA-WEEDS.

SEA and land are, after all, wonderfully like each other. The surface of the land has its mountains, its valleys, its fire-vomiting volcanoes, its mountains of eternal cold. So the bed of the sea is delved into vast valleys, as yet unfathomable by human plummet; and these valleys we of the upper world call depths. Also, it has its precipitous mountains, some towering above the watery surface, and others lifting their heads until they are dangerous neighbours to those that go in ships upon the waters; and these we call by various insulting names according to their degree of elevation. And there are volcanoes of the sea as well as of the land; while the Polar islands, which are, in fact, the tops of submarine mountains, are covered with snows as eternal as those which crown the Monarch of mountains himself.

Then, the sea-bed has its Table Mountains, its vast Saharas, its undulating prairies, its luxuriant forests, and its verdant pasture-lands. And as the sandy tracts or shingly beds are bare and devoid of vegetative life on the upper earth, so are they also in the sea below; while submarine forests lift their branches towards the light of the sun, and submarine herbage waves its many-coloured leaves in the rolling sea, just as flowers and leaves bend to the breezes above. For in the kingdom of Ocean, water is the atmosphere, and, like its more ethereal relative, is ever rolling, and ever changing.

Let us now visit the boundary line of the two great **kingdoms**, Earth and Water, and though belonging **to** the former, extend our researches as far as possible into the latter.

Throughout the preceding pages it will be noticed that the expression "at low water" is constantly used. Now, this expression is quite necessary; for were the sea always to remain at the same height, our knowledge of its wonders would be wofully circumscribed. It is little enough even now, but that little would be almost reduced to nothing were there no alternations of high and low water.

Of the theory of tides there is here no opportunity to speak, for it is a most complex subject, and even to give a hasty sketch would occupy many pages, and require many diagrams. Suffice it to say, that the grand exciting cause of the tides is the force called attraction, or gravitation; the moon being the chief among the many agents through which it acts. It matters not whether the water is salt or fresh, whether as an ocean it fills the bed of the Atlantic, or as a drop of dew trembles on a violet leaf. The tide-force still acts on it, and tides there are, although we are incapable of perceiving them. It is the same with the upper sea, namely, the atmospheric air of our earth. In the aërial ocean there are waves, whirlpools, calms, and storms, although our eyes are too dull to perceive them, and can only be made aware of their existence by seeing their effects.

Twice in every day of twenty-four hours the water advances and recedes, and thus at least one opportunity is given daily for the observer to follow the retiring waves, and to discover some small portion of the wonders of the sea. Some of its living and breathing inhabitants have been mentioned in the preceding chapter, and in the following pages will be briefly described some few of its vegetative inhabitants, that breathe not, but yet live.

If we walk on the sea-shore, vast masses of dark olive vegetation meet our eyes; if we wait until the tide has retreated, and examine the pools of water that are left among the rocks, there we find miniature forests, and gardens of gorgeous foliage, some of which are scarlet, others pink, others bright green, others purple, while some there are that play with all the prismatic colours, each leaf a rainbow in itself. If we take a boat, and rowing well out to sea, cast overboard a hooked drag, we shall find adhering to the iron claws new kinds of vegetation, and probably among them will be found a veritable flowering plant,—apparently as much out of its place at the bottom of the sea as a codfish in a birdcage. Now all this luxuriant, graceful, and magnificent foliage, we dedecorate with the title of sea-*weed*. It is a miserable appellation; but as it is a term in general use, I shall employ it, although under protest.

Those sea-weeds, then, which first strike our eyes, are usually those denominated Wracks, the Common Bladder-wrack (*Fucus vesiculosus*) being the most common. For a figure of this plant, see plate J, fig 6.

There is little difficulty in distinguishing this conspicuous alga; for the double series of round air-vessels with which the fronds are studded, and the mid-rib running up the centre of each frond, point it out at once. This plant, together with one or two others of the same genus, is still used in the manufacture of kelp, but not to such an extent as was formerly the case. There is a variety of this plant found in salt marshes, where it congregates in dense masses: this variety is very small, being only an inch or two in height, and the eighth of an inch, or even less, in width. The plant is at all times very variable, according to its locality, both in colour and form.

When trodden on, or otherwise suddenly compressed, the air-vessels explode with a slight report, and seem to afford much gratification to juveniles. This and other fuci grow in the greatest abundance on rocks that are covered by the waves at high water, and left bare when the tide retires. Now on, under, and among these rocks, the great zoological or botanical harvest is to be collected, and therefore among these rocks the collector must walk.

I make mention of this circumstance, because it is necessary to warn the enthusiastic but inexperienced naturalist, that the slimy and slippery fuci make the rock-walking exceedingly dangerous; for the masses of fuci are so heavy and thick, that they veil many a deep hollow, or slightly cover many a sharp point,—in the former of which a limb may be easily broken, and by the latter a serious wound inflicted,—and there is special reason for avoiding any such mishap. Proverbially, time and tide wait for no man; and should a disabling accident occur when no one was near to help, the returning waters would bring death in their train—a death the more terrible from its slow but relentless advance.

Now, the reader must be careful not to confound with *Fucus vesiculosus* another species of somewhat similar appearance, namely *Fucus nodosus*; see plate J, fig. 1.

This plant may at once be distinguished from the Common Bladder-wrack, by the absence of a mid-rib; it is of a tough consistence, and it grows to a large size, being sometimes nearly six feet in length.

About half-way between high and low water another species of fucus may be found: this is destitute of air-vessels, it lacks the sliminess of the bladder-wrack, and its edges are toothed, like the edge of a saw. It is much about the same size as the bladder-wrack, but perhaps rather longer; see plate D, fig. 2.

This is a very useful plant indeed. It is a capital manure for land, it can be preserved, and used as food for cattle, it can be made into kelp, and it is an excellent substance in which to pack lobsters, and other marine productions that are sent inland. The bladder-wrack is much used for the same purpose, but its sliminess renders it liable to heat and to ferment; while

Fucus serratus, being comparatively free from slime, retains its cool dampness, and preserves the fish sweet. It is really of importance, for a tainted lobster is not only nauseous to the palate, but even dangerous to the whole system.

There is a very tiny fucus, some four or six inches long at the most, that is to be found near high water-mark, chiefly in summer and autumn; it may be recognised by a number of small channels that furrow one side of the frond. It has no air-vessels.

All these plants, together with all the algæ comprised in this chapter, belong to the class of algæ called MELANOSPERMS, or black-seeded; so called from the dark olive tint of the spores, or tiny seeds, from which they spring. They all seem to be exclusively marine.

At spring-tides the waters recede considerably below their usual mark, and these seasons are the harvest-times of the shore-naturalist. As nearly as possible at six hours after the high tide the waters will have retired to their lowest boundary, and near that boundary will be found myriads of new forms, both animal and vegetable. Indeed, so prolific is the spring-tide harvest, that an hour or two of careful investigation will sometimes produce as good results as several hours' hard work with a dredge. It is better to go down to the shore about half an hour or so before the lowest tide, so as to follow the receding waters, and to save time.

When the naturalist has gained the spots below the usual low-water mark, he will find himself in the midst of a new set of vegetation, contrasting as strongly with the productions of the higher grounds, as forest trees with herbage and brushwood. Huge plants, measuring some eleven feet or so in length, and nearly a yard in width, are firmly anchored among the rocks by roots rivalling in comparative size and strength those of the oak-tree. This plant is commonly known by the name of Oar weed, and may be easily recognised from the drawing in plate D, fig. 1. Its scientific name is *Lami-*

naria digitata. It is called "Laminaria" on account of the flat thin plates, or laminæ, of the frond, and "digitata," or fingered, because the frond is split into segments, something like the fingers of a hand.

I may as well mention here, that the sea-weeds have no real root, and do not derive their nourishment from the soil, as do the plants of earth; they adhere to the rocks or stones by simple discs, and draw their whole subsistence from the water that surrounds and sustains them. In the so-called root of the Laminaria there are no root-fibres, but a succession of discs, each connected with the main stem of the plant by a woody cable.

The stem of the Laminaria is very strong, and is used for making handles to knives and other implements. When fresh, this stem is soft enough to permit the tang of a knife-blade to be thrust longitudinally into it. A portion of the stem sufficiently long for the knife-handle is cut off, and in a few months it dries, contracting with such force as to fix the blade immovably; and having much the consistency and appearance of stag's horn. One good stem will furnish more than a dozen of these handles.

Among the Laminariæ may be seen growing a singular plant, more like a rope than a vegetable. It consists of one long, cylindrical, tubular frond, hardly thicker than an ordinary pin at the base, but swelling to the size of a swan's quill in the centre. When the plant is handled, it slips from the grasp as if it were oiled; this effect is produced by a natural sliminess, aided by a dense covering of very fine hairs.

The name of this plant is *Chorda filum.* Its length varies extremely, some specimens being found to measure barely one foot, while others run from twenty to thirty, and even to forty feet. It tapers gradually from the middle to the point, where it is about the same thickness as at the base.

I here make an exception to my general rule of

excluding all but the commonest objects, in favour of one sea-weed, which, although not very common, yet may be found quite unexpectedly. It owes its introduction to its very singular form. The name of it is the Peacock's-Tail, deriving its title from its shape. Its scientific name is *Padina pavonia;* see plate A, fig. 3.

The habitation of this plant is midway between high and low water-mark, where it may occasionally be found adhering to the rocks. It is not a large plant, as it is generally only two or three inches in height, but occasionally reaches the height of five inches.

In the same order as the Padina is another little algæ, which, I think, is one of the prettiest of the Melanosperms. I do not know whether it possesses any popular name, but its scientific title is *Dictyota dichotoma*. For a figure of it, see plate A, fig. 5. It is a very delicate-looking plant, and, unlike the Melanosperms in general, lives tolerably well in an aquarium. The name Dictyota is derived from a Greek word, signifying a net; and it will be seen, on examination, that the surface of the frond appears as if woven into a tiny network, with square, or rather slightly oblong meshes. Its specific appellation of "dichotoma" is also of Greek derivation, and signifies "cut in pairs," in allusion to the shape of the frond.

Failing space permits only one more plant belonging to the class, or rather, to speak accurately, the sub-class Melanospermeæ. This is the plant known to botanists by the title of *Ectocarpus siliculosus*, and which I mention here because it is liable to be confused with other algæ that much resemble it in form, though not in constitution; see plate A, fig. 2. It is called Ectocarpus from two Greek words, signifying "external fruit," and its specific title "siliculosus" is given to it on account of the silicules, or little pod-like bodies, that are found on the branches. These details are of very minute size, and cannot be made out without the assistance of a magnifying glass.

CHANGE OF TINT.

These dark-spored vegetables are very variable in colour, as indeed are all the algæ, without reference to the colour of their spores; sometimes, indeed, even trespassing on the colour of another sub-class. These changes are mostly due to the varying depths of the sea where the plants grow, and to the amount of light and shade which falls to their lot. Even the hardy, rough, and coarse bladder-wrack, which is usually of a very dark olive-green, more approaching to black than to green, becomes of a rich yellow tint when found at any depth of water.

When dried the green vanishes totally, the colour changing to dark-brown, and in many cases to black. Most of this sub-class of algæ require alternations of water and air, the best specimens being found where they are exposed to the heat of the sun and to the force of the winds for some hours daily.

CHAPTER IV.

RED-SPORED AND GREEN-SPORED ALGÆ.

With this chapter we begin the account of another sub-class of algæ, the Rhodosperms, or red-seeded. The plants belonging to this class are among the most beautiful of the algæ, that is, when they are placed in favourable situations; for they also change their colours, and as their most beautiful colour is their natural tint, any change is for the worse. Some of them even become brown when there is too much light for them.

About low water-mark may be found growing largish masses of a dense, thread-like, reddish foliage, sometimes adhering to the rock, or sometimes even fixed to the stems of the Laminaria. When removed from the water the plant does not collapse, like many of its relatives, but each thread and branch preserves its own individuality. This is one of the large genus Polysiphonia, and the specific name is "urceolata." See plate K, fig. 2.

By the side of the plant itself is represented a little object that explains the latter title. This little jar-shaped object is one of the fruits, or ceramidia, as they are learnedly called, much magnified. The word "urceolata" signifies pitchered, if we may be permitted to coin an English word corresponding to the Latin. The name Polysiphonia is Greek, and signifies "many siphons," or tubes. The reason for the name is evident on cutting any of the branches transversely. It will be then seen that the plant is composed of six tubes arranged round a central aperture; the branches are jointed, the length of each joint being several times its own width.

There are twenty-six known British species of this single genus.

That popular author and extensive traveller, Baron Munchausen, tells us that in one of his journeys he met with a tree that bore a fruit filled interiorly with the best of gin. Had he travelled along our own sea-coasts, or indeed along any sea-coasts, and inspected the vegetation of the waves there, he would have found a plant that might have furnished him with the ground-work of a story respecting a jointed tree, composed of wine-bottles, each joint being a separate bottle, filled with claret. It is true that the plant is not very large, as it seldom exceeds nine or ten inches in height; but if examined through a microscope, it might be enlarged to any convenient size.

The name of this plant is rather a long one, but very appropriate, *Chylocladia articulata*, i. e. the "jointed juice-branch." See plate A, fig. 1.

It may be found adhering to rocks, or sometimes parasitically depending on some of the larger algæ, and really does resemble a jointed series of transparent bottles filled with claret or other red wine. The colour is remarkably delicate and beautiful, but is rather apt to fade after a time; when it is preserved, dried, and pressed, the gelatinous juice that filled the interior disappears, and the plant can be flattened until it hardly presents any thickness, even to the touch. There is now before me a dried specimen of another species of chylocladia, which adheres so firmly to the paper on which it is laid, and is so delicate in substance, that several persons to whom I have shown it have mistaken it for a well-executed drawing.

If now the reader will refer to plate C, fig. 1, he will there see depicted one of the most remarkable of the algæ; remarkable in itself, and for the great battles which have been fought over it by scientific individuals. This plant is the Common Coralline (*Corallina officinalis*), which may be found most abundantly on any of our coasts, growing in greatest perfection near low watermark.

It is well enough known that many creatures, formerly supposed to be vegetable, such as the corals and the zoophytes, have since found their proper place in the animal kingdom; and one consequence of this reformation was, that several real plants were supposed to be animals, because they possessed some of the characteristics which had distinguished those animals that had been placed in their proper position. Of these plants the coralline is a good example; for until a comparatively late period, it was placed among the animals in company with the true corals.

There was reason for this error, for the coralline is a very curious plant indeed, gathering from the seawater, and depositing in its own substance, so large an amount of carbonate of lime, that when the purely vegetable part of the alga dies, and is decomposed, the chalky portion remains, retaining the same shape as the entire plant, and very much resembling those zoophytes with which it has been confounded. While growing, the coralline is of a dark purple colour; but when removed from the water, the purple tint vanishes, and the white stony skeleton remains. It is, however, a true vegetable, as may be seen by dissolving away the chalky portions in acid: there is then left a vegetable framework, precisely like that of other algæ belonging to the same sub-class.

The coralline is a small plant, seldom exceeding five or six inches in height, and not often even reaching that size. However, it compensates for its low stature by its luxuriant growth, being usually found in dense masses wherever it can find a convenient shelter.

If a dried branch of coralline be inserted into the flame of a candle, it exhibits a most brilliant white light just at the point where it meets the flame. The light is exhibited better by the flame of a spirit-lamp than by that of a candle, and for obvious reasons.

It will live well in an aquarium, and, if tastefully disposed, is an elegant ornament to the vase or tank. There is now in my own aquarium a moderate tuft of coralline, which seems in good health, although the water has lately been assuming an unpleasant milky appearance, from some cause which I cannot as yet detect.

We now come to a most magnificent sea-plant, magnificent both on account of its gorgeous colouring, and on account of its luxuriance. This is the *Delesseria sanguinea*, represented, about half its usual size, in plate J, fig. 3.

The shape of the leaf, or rather of the frond, so closely resembles that of terrestrial trees, that at first sight few would attribute the beautiful scarlet leaf, with its decided midrib and bold nervures, to an alga. Yet an alga it is, and may be found in its most perfect state about June or July: later in the year it becomes very ragged, the broad flat frond giving way to the fruit. In this state, although interesting to the botanist, it is hardly suitable for the cabinet, as little of the plant is left except the midrib, and a few flapping raglets. When spread on paper and preserved, it retains its colour well, and adheres very firmly.

The fronds are generally from two to seven or eight inches in length, but they are not often found exceeding five or six inches. A branch containing eight or ten fronds, averaging five inches in length, may be considered a good specimen, and worth preserving, if the edges are entire. There is a very peculiar marine scent about this plant, an "ancient and fishlike smell," quite indescribable, but not to be forgotten. A large branch will retain this scent for months. I have by me a tuft of this plant, which I gathered in July last, and its peculiar smell is now (April) very perceptible.

There are five British species of this beautiful genus, none of them very rare. *Delesseria hypoglossum* (plate D, fig. 4) may be found in the summer months growing on almost every coast. It is a very pretty plant, although not so gorgeous as its predecessor. The fronds are generally of small size, being hardly a quarter of an inch in length.

In the little sea-weed landscapes, that are sold so abundantly at the fashionable sea-side towns, there is one species of sea-weed in great request for trees and bushes. It is of a bright pinky red colour, and is thickly branched, so as to afford a tolerable representation of a forest tree, or of a thick bush. This is the *Plocamium coccineum*, a plant sufficiently beautiful to the unassisted eye, but especially so when submitted to a magnifying lens. When examined through a glass of moderate power, it will be seen that even the tiny branchlets, each hardly thicker than a hair, are again furnished with a row of smaller ramifications, somewhat resembling a very finely-toothed comb.

On plate K, fig. 3, may be seen a specimen of the Plocamium of the natural size, and near it a single branch magnified, in order to show the tiny combs.

Many of the marine algæ are used as articles of food; some eaten uncooked, and others after a long course of boiling. To the former of these categories belongs the Dulse, Dillisk, or Dillosk (*Rhodymenia palmata*), although it is sometimes cooked. The species, however, which is here illustrated, is *Rhodymenia bifida*, a plant of a very fine rosy red when fresh, found in tolerable profusion adhering to rocks or on the larger algæ. The fronds are generally two inches or so in length, and about a quarter of an inch in width. For a figure of this plant, see plate K. fig. 4.

The Carrageen Moss, so well known in the form of jelly and size, is one of the Rhodosperm Algæ, by name *Chondrus crispus*. (Plate J, fig. 5.)

It may be found growing on the rocks in large quantities, where its shape will be the best guide to its detection, for its colour is exceedingly variable. Although one of the Rhodosperms, it is very frequently of a greenish tint, and in many places it assumes a yellow jaundiced complexion, not at all of a healthy nature.

To preserve it for esculent purposes, it must be washed in fresh water and then left to dry, when it soon becomes horny to the touch, and resists pressure. If boiled, it subsides into a thick colourless jelly, that is thought to be very nutritive, and is employed for many purposes. Invalids take it in their tea, or epicures in their blanc-mange. Calico printers boil it down into size, and use it in their manufactures. It is said to be a good fattening substance for calves, if boiled in milk; and, lastly, pigs are very fond of it when it is mixed with potatoes or meal. It is sometimes known by the name of "Irish Moss." It will grow in an aquarium.

A plant is represented on plate J, fig. 4, that is found plentifully between tide-marks. It is rather a conspicuous plant, and is appropriately named *Furcellaria fastigiata*, the generic title being derived from a Latin word signifying a little fork. It is of a dark-brown colour with an obscure dash of purple, but in drying the purple departs, and the brown becomes nearly black.

I have already mentioned that some of the algæ reflect prismatic colours. This is occasionally the case with *Chondrus crispus*, and there is one genus which is so resplendent that the name *Iridæa* is given to it; Iris signifying a rainbow. The species represented at plate C, fig. 4, is *Iridæa edulis*, a plant which is sometimes eaten raw, and sometimes fried by unpoetical gastronomists. I do believe that some people would fry the rainbow itself if it were eatable.

The frond of this species is generally about nine or ten inches in length, and five inches in width, although it sometimes nearly doubles these dimensions. Its colour is an uniform deep red, and its shape somewhat resembling a battledore.

A particularly elegant species of alga, making a good figure when spread on paper, is seen figured on plate K, fig. 5. The fronds are sometimes more than a foot in length, but do not often exceed ten or eleven inches, some being only three or four inches long. The colour is rather apt to fly, unless care be taken; but it is a beautiful plant, were it only for the elegance of its form. Its name is *Ptilota plumosa*, both words having a like signification, and meaning "winged," or feathery.

There is a pretty little alga, called *Griffithsia setacea*, which has the property of staining paper with a fine pinkish-scarlet hue, when the enclosing membrane bursts. Contact with fresh water will usually cause the membrane to yield, and then the colouring matter is shot out with a slight crackling noise.

Its length is generally about four or five inches. A drawing of the plant of the natural size, together with a magnified sketch of the fruit, may be seen on plate K, fig. 1.

The last of the Rhodosperms that will be noticed in this volume is a very delicate species, entitled *Nitophyllum punctatum;* see plate C, fig. 5. This plant will easily be recognised from the drawing. Its usual size is six or ten inches in length, and nearly as wide; but it is not uncommon to find specimens that exceed a foot in length, while some huge monsters have been found that measured five feet in length and a yard in width. It is easy enough to distinguish this plant from the Delesseria, as it has no midrib.

The CHLOROSPERMS, or Green-seeded Alga, are the best friends of those who keep marine aquaria, for they are endowed with the power of pouring out oxygen in very large quantities when placed in favourable circumstances. If any of my readers wish to preserve alive the creatures that they find on the sea-shore, they can do so without difficulty, by imitating as nearly as possible the natural state and accompaniments of the animals which they have captured.

If even one or two fish, crabs, or, indeed, any living animals, be placed in a jar of sea-water, they speedily exhaust the free oxygen of the water, and, as the water cannot absorb fresh oxygen from the atmosphere so rapidly as the animals consume it, the water soon becomes unfit to support animal life, and its inhabitants die as surely as a man would who was enclosed in an air-tight box. It is possible to renew the oxygen by dashing water into the jar from a height, or even by pumping fresh air into it; but such a process would be very fatiguing, as it must be continually carried on day and night. But it is found that plants have the property of pouring out oxygen when they are in a healthy state and acted on by light. So, if we can procure plants that will thrive in a confined space, and keep them in a light room, we shall find that each plant acts as a natural pump, and not only supplies continually fresh oxygen, but consumes the carbonic acid gas that loads the water with its stifling influence. The Chlorosperms are peculiarly useful for this purpose, as many of them will live for an unlimited time in confinement, continually regenerating the water in which they are placed. I have now an aquarium containing water that I brought from the sea last August, and by the untiring exertions of a few green sea-weeds the water has been preserved bright and pure, even though inhabited by all kinds of marine animals.

Among the most useful, as well as the most elegant of the sea-weeds used for this purpose, is the little *Bryopsis plumosa;* see plate D, fig. 3. This brilliant and delicate little plant is common enough, and may be found in the pools left by the retiring tide, where it adheres to their rocky walls. The colour of the plant is a very bright green, and its form is so feathery, or rather fan-like, that it well deserves its name of "plumosa."

In almost any little pool, between tide-marks, or even hanging from rocks that have been left quite dry, may be seen thick tufts of a coarsish horse-hair-like plant, of a dull green colour, often dashed with black. This is the *Cladophora rupestris*, one of the commonest species out of the twenty that are exclusively marine. There are two species that inhabit ditches and lakes where the sea occasionally obtains admission, and several others that prefer water entirely fresh. The length of the tufts is about four or five inches, often less, but seldom **more**.

Another species of the same genus, *Cladophora arcta*, is of a brighter green than the preceding, and altogether a prettier plant. It grows in a radiating manner from a very broad disc. This plant is represented on plate c, fig. 2.

But the most useful of the Chlorosperms may be found almost at the very margin of high water, where they live rather more in the open air than under water. These are the *Ulvæ* and *Enteromorphæ*, the first being known by the popular title of Laver, and the second of Sea-grass. There is another plant that is also called Sea-grass; but it is not an alga, and will be mentioned at the end of this chapter.

The Common Sea-grass (*Enteromorpha compressa*) may be seen in abundance on the stones and rocks that are even for a few hours submerged daily. The leaf, or rather frond, of this species is variable in width, sometimes being hardly wider than common sewing thread, and sometimes so wide as to resemble a very narrow ulva. It is this variety which is represented in the engraving, plate c, fig. 3. When the waves retire, leaving sundry pools fringed with this and other seaweeds, their fronds form hiding-places for innumerable living beings of very many species; and by gathering masses of the wet weed into a basket, and then putting it into a large vessel filled with sea-water, myriads of animals may be captured with hardly any trouble. They will live perfectly well in the vessel if it is kept in a light spot with a free circulation of air.

The Common Green Laver (*Ulva latissima*), plate K, fig. 6, sometimes called the Sea Lettuce, is found most abundantly on the same spots as the preceding plant. Of all the sea-weeds for an aquarium, the Green Laver is perhaps the very best. It is very pretty, from its delicate green colour, and the various folds and puckers into which it throws itself. Its power of expiring oxygen seems to be almost unlimited. I have in my aquarium a large plant of this species, which generally lives very contentedly in the place where it had been deposited. But, a few days ago, the sun shone brightly enough to pierce through the veil of smoke with which the metropolis is generally hidden from his presence, and consequently there was a greater abundance of light than usual. On looking at the aquarium, I found that the ulva had risen in the water, and was hanging in most elegant festoons from the surface, forming emerald caves and grottos such as the sea-nymphs would love. Even at a little distance it was a pretty sight, but a closer inspection revealed still more beauties; for being excited by the unwonted light, the plant had poured forth so much oxygen that its entire surface was thickly studded with tiny sparkling beads, that had buoyed up the whole plant, each bubble acting as a miniature balloon. When, however, a black cloud came over the sun, the bubbles soon detached themselves, ascended to the surface, and as there were no more to take their place, down dropped the plant to the bottom.

On a bright day the little oxygen bubbles are so rapidly exuded, that they quite fill the water, rising to the surface, and there dissipating, very much like the sparkling air-bells in champagne.

The Purple Laver, as it is called in England, or the Sloke, as it is termed in Ireland, is another of these useful plants. In external appearance it very much resembles the ulva, save that the colour, instead of being light green, is purple. From this peculiarity of colour it is called by botanists, Porphyra, which word signifies "purple." There appear to be only two species of this genus belonging to our coasts, the one *P. laciniata*, and the other *P. vulgaris*. The former of these plants is engraved in plate A, fig. 4. Only a portion of the frond is given. It is hardly inferior to the preceding plant in value to the aquarium-keeper, and flourishes wonderfully. *P. vulgaris* may be used for the same purpose. I have seen one of these plants in an aquarium, which had increased to such an enormous size, that it was aptly compared by a bystander to a lady's purple silk apron.

The ulva and porphyra, if intended to be eaten, must be gathered in the winter, or, at all events, the very earliest of the spring months. The purple laver is said to be much superior to its green companion, but I cannot speak from personal experience. If any of my readers would like to try the experiment for themselves, they may easily do so; the laver should be stewed for several hours, until it is reduced to a pulpy mass, which, with the addition of lemon juice, is considered by some persons a dainty.

I may here mention that, although both ulva and porphyra will live in an aquarium, when floating freely through the water without any attachment, yet it is better that they should be adherent to some stone or shell, by which they can be anchored in a convenient spot. Now these plants are very constant, for they never have but one attachment during their whole lives, and if torn from that one object they never affix themselves to any other: so it is necessary to use a chisel and mallet, or at all events a geologist's hammer, for the purpose of detaching the portion of rock or stone to which the plant is adherent. Generally the geologist's hammer, if properly chosen, answers every purpose. Almost at the commencement of my last shore-season I dropped both my chisels into a rock-pool, and not being able to find them again, brought the hammer into play; and so useful was that hammer, that I did not find it requisite to procure a fresh set of chizels during the four weeks of shore-searching.

A plant has been mentioned, which does not belong to the sea-weeds, although from its residence at the bottom of the sea it is often thought to be of that family; this is the *Zostera marina*, for a drawing of which see plate 5, fig. 2, a true flowering plant, growing with a real root at the bottom of the sea. Its entire character is so completely terrestrial, that it can at once be distinguished from the alga.

The zostera is an useful plant to the zoologist, for it grows in great numbers, or rather in great fields, affording pasture to innumerable living beings, which he captures in his net or dredge. It will live well enough in an aquarium, and gives a decided character to that portion of the tank in which it is placed. Again, when dried, it is largely manufactured into bed-stuffing, under the name of Alva, and is used instead of hay or straw for packing glass, china, and other fragile wares. On many coasts this plant is known by the name of Grass-wrack, and is cast up in great quantities on the shore, where it soon turns black, rots, and presents a very unsightly aspect.

If the naturalist wishes to dry and preserve the algæ which he finds, he may generally do so without much difficulty, although some plants give much more trouble than others. It is necessary that they should be well washed in fresh water, in order to get rid of the salt, which being deliquescent, would attract the moisture on a damp day, or in a damp situation, and soon ruin the entire collection. When they are thoroughly washed the finest specimens should be separated from the rest, and placed in a wide shallow vessel, filled with clean fresh water. Portions of white card, cut to the requisite size, should then be slipped under the specimens, which can be readily arranged as they float over the immersed card. The fingers alone ought to answer every purpose, but a camel's-hair brush and a needle will often be useful. When the specimen is properly arranged, the card is lifted from the water, carrying upon it the piece of sea-weed.

There is little difficulty in getting the plants to adhere to the paper, as most of the algæ are furnished with a gelatinous substance, which acts like glue, and fixes them firmly down. Where they do not readily adhere, the use of hot water will generally compel them to do so; and if they still remain obstinate, the gelatine obtained by boiling the carrageen (*Chondrus crispus*)—see p. 39—will be an unfailing remedy. This is a much better cement than animal glue, or even gum-water, as it approaches nearer to the natural glue of the plant. *Furcellaria fastigiata*, *Cladophora arcta*, and others, are not easily affixed to the paper, and will often require the aid of some adventitious substance.

This sketch of the British marine algæ is necessarily very imperfect; but even as it is, they have occupied rather more than their proper share of paper. Still, there are sufficient different genera here mentioned to prevent the inexperienced marine botanist from erring very widely, and the specimens chosen have been selected for two reasons; the one being that they may be found on almost every coast; and the other, that they form a series of landmarks, by means of which the observer can be directed in the right course.

CHAPTER V.

EGGS OF MARINE ANIMALS—CUTTLES AND THEIR HABITS.

It is impossible to walk on the sea-shore without being struck by the strangely-shaped objects that are cast up by the waves, and left high and dry until swept away by the next tide, which in its turn brings new and varied forms; and these objects are continually changing according to the season of the year. One week may pass, and the observer will see nothing on the sand with which he is not thoroughly acquainted; and in the course of the next week new and grotesque objects will be found profusely scattered at his feet. Some of these objects are purely natural, and their presence is occasioned by the development of nature, while others are but a mixture of the natural and artificial.

For example, when green peas come into general use, the empty pods are thrown into the sea, and are after a space washed up by the waves, having been so chemically acted upon by the salt water, so abraded by sand and pebbles, and so nibbled by various marine animals, that they can hardly be recognised even by the very persons who have consumed the peas that were once enshrined in these metamorphosed husks. Nutshells, gooseberry husks, currant stalks, cherry stones, and many similar objects, assume, after a temporary sojourn in the ocean, very singular forms, and may easily deceive an unaccustomed eye, especially as they often resemble vegetable and animal remains that properly belong to the sea. I mention this, because I have seen many instances of such deception.

Only a year or two ago, I commissioned a friend to procure for me any marine curiosities that could be found, and to forward them when a sufficient quantity had been amassed. In due time the parcel arrived, having a pleasant marine smell about it, and on being opened, was found to contain some very curious objects. Among them was one on which the collector especially prided himself, and in chase of which he had bravely waded into the sea and undergone a complete wetting. It was apparently a kind of sponge, about eight inches in length, of a light brownish yellow colour, and hollow at one end, as are most sponges. But, on a nearer examination, this sponge proved to be a cabbage-stalk, of which only the fibrous portion remained, and which had probably been tossing about for many months in the sea; sometimes soaked by the waters, sometimes lying on a rock and bleaching in the sun, until the next high tide carried it back again; and at other times entangled among heavy sea-weeds, and anchored by them under water. I still preserve it as an example of marine curiosities.

It is quite necessary, therefore, to exercise much caution in selecting objects. The surest mode of obtaining success is to gather indiscriminately everything that presents itself, and having conveyed the cargo to a place of shelter, deliberately to examine the heap. By so doing, the valuable objects will be retained, and the useless rejected, without so much danger of passing over the one or preserving the other, as if the choice were made immediately. After a time, the eye will become so accustomed to note distinguishing characters, that such a process will be no longer required, and the eye will make its selection at once.

Among the singularly shaped substances that are found thrown on the sands, are the eggs of various marine creatures. Many of these eggs are so curiously formed, that they would hardly be recognised as such by one who was not acquainted with the animals to which they belonged. Very many eggs are found on the shores; but as most of them may be referred to one out of four or five classes of animals, I will only mention those that are, as it were, the general types.

Plate H is specially dedicated to eggs, and, as will be seen on referring to the plate, some of them have anything but an egg-like aspect. The commonest of all the eggs, masses of which are to be found at almost all seasons of the year, are those of the common whelk, the shell of which is represented on p. 16, and its eggs on plate H, fig. 3. The egg mass from which the drawing was made, is now before me. It contains many eggs as yet unhatched, many which are addled, and some which have already discharged their inhabitants. It was taken out of the sea at the very beginning of April, but if it had been permitted to remain in its habitation until the summer, all the egg-sacs would have been found empty. I am keeping it in the aquarium in faint hopes that some of the young whelks will be hatched; but it is very doubtful whether the surrounding conditions are sufficiently favourable. The enormous size of some of these egg-clusters is remarkable; for the whelk itself is by no means a large shell, and so it often happens that those persons who are practically acquainted with the whelk, but not with the eggs, entirely refuse to believe that there is any connexion between objects so dissimilar. The empty egg-cluster bears some resemblance to a rather dingy honey-comb, partially squeezed between the hands. But when the membranous egg-sacs contain their living inhabitants, the animal nature is evident from the presence of the young whelks, whose forms can be plainly seen through the semi-transparent substance that envelopes them until they are sufficiently strong to lead an independent life.

A description of the purpura has already been given on p. 17, and it will be seen that this creature is interesting, not only on account of the beautiful dye which it contains, but also for the singular shape of its eggs, a cluster of which is represented in plate H, fig. 2. Sometimes these curious eggs are found affixed to little stones, and, indeed, when first deposited, some of them seem always to be thus anchored as it were, and to afford support to the others, who stand on each other's shoulders, something like the human pyramid that is occasionally formed at Astley's and similar establishments. The cluster from which this sketch was made I found lying among the rocks, and put it carefully away. But in the course of travel, the box in which it was placed gave way, and my poor little egg-cluster was thrown among a large boxful of shells. There it rested for some seven or eight months, and when discovered was dry, shrivelled, and hardly to be recognised. But, when placed in hot water, it absorbed the liquid as if it had been composed of blotting paper, and in two minutes had completely resumed its natural aspect.

Often may be found, lying on the shore, masses of dark soft substances, not unlike purple grapes, both in size and in shape. If you ask a fisherman the name of them, he will tell you that they are sea-grapes, but for any further information you may usually ask in vain. Indeed, as a general fact, those who live on the seashore are hopelessly ignorant of its treasures. I knew a person of intellect, education, and ordinary observing powers, who had resided within a stone's throw of the sea for a period of thirty years, who had been accustomed to walk on the sands almost daily, and yet had never in his life seen, and hardly ever heard of, a common sea anemone, although the shores were studded with them as the sky with stars. And one of those strange amphibious humanities, who get their living by collecting shells, curious pebbles, sea-weed, zoophytes, and other saleable curiosities, persisted in declaring that the hermit crab was the young of the common edible crab, and that when it grew old enough, and was too large for its shell, it abandoned the useless adjunct, and commenced another course of life.

But to return to our sea-grapes, of which a sketch may be seen on plate H. fig. 5. These are the eggs of a cuttle-fish, and curious eggs they are. Each is produced into a flexible stalk, by means of which the mass is held together, and affixed to any convenient object. The egg-cluster from which the sketch was taken was one of four or five which I preserved at different times, in order to watch their progress. Here and there, among the dark mass of eggs, appeared one nearly white, and semi-transparent, through whose delicate walls might be seen the little cuttle within, very lively and seemingly anxious for his emancipation. At the bottom of the egg-cluster may be seen one of the young creatures escaping from the prison that had confined him, and, as will be seen, the young cuttle is rather a comical-looking little animal.

CUTTLE.

I was much amused with the perfect self-possession of the first that was hatched in my presence. It had not been free from the egg-shell for one minute before it began a leisurely tour of the vessel in which it first saw the light, examining it on all sides, as if to find out what kind of a place the world was, after all. It then rose and sank many times in succession over different spots, and after balancing itself for a moment or two over one especial patch of sand, blew out a round hole in the sand, into which it lowered itself, and there lay quite at its ease. It executed this movement with as much address, as if it had practised the art for twenty years.

The mode by which the creature forms this little burrow is sufficiently curious. Its siphon is a slightly projecting tube, and by bending this towards the spot selected, and then forcibly ejecting a column of water, the sand is displaced apparently by magic. This siphon is also useful as a means of progression; and, in one of the cephalopods, as these creatures are called, because their feet are situated on their head, has not only cast out water, but also the mistaken notions with which careless observers had obscured its history.

Some cephalopods have bodies soft and naked, while others are protected by a shell secreted by themselves. Among these shelly cuttle-fish is the well-known nautilus, who was once said to row himself over the sea with his legs, and to stretch out his wing-like arms as sails to catch the breeze. But it is now known that these sails are kept closely wrapped round the shell, which, indeed, they secrete originally, and can mend if injured, while the legs are suffered to trail loosely. Successive jets of water are then ejected from the siphon, by which the creature is driven in a contrary direction. By this water-power the nautilus is urged through the waves, but when it wishes to move about on the bottom of the sea, it just crawls exactly as a large spider might be supposed to do.

On the arms, legs, feet, or tentacles of the cuttles, are arranged rows of suckers, which are capable of taking a very firm hold of any object to which they are applied, aided in some species by sharp hooks. If any one of these suckers is examined, it will be found to be the living type of the air-pump, an exhausting syringe that was in full operation thousands of ages before man worked in metals, and more perfect than the best air-pump ever made. For there are extant many specimens of fossil cuttle-fish, the relics of one of which, by the way, are familiar to most people by the title of "thunderbolts"—long cylindrical bodies, composed of calcareous spar, pointed at one end, and slightly hollowed at the other, not unlike an elongated Minié bullet, and which, when cut or broken across, display a radiated structure.

If an arm of a cuttle be taken, and any one sucker examined, it will be seen to consist of a thick muscular membranous cup, having a cavity at the bottom, something like the chamber at the bottom of a mortar. The sucker should now be divided longitudinally, and then at the end of this "chamber" will be seen a soft muscular piston, exactly fitting the cavity. Now, if the circumference of the sucker be closely pressed against any substance of sufficient size and consistency, and the piston withdrawn, a vacuum is at once created, and powerful adhesion takes place. As, on an average, each cuttle is furnished with nine hundred suckers, the force of its hold may be imagined.

There is a substance that is often to be picked up on the shore, and oftener to be purchased at the perfumer's shops, known by the name of cuttle-bone, and when reduced to powder, used for various purposes. This so-called cuttle-bone is not bone at all, but a very wonderful structure, consisting almost entirely of pure chalk, and having been at one time embedded loosely in the substance of some departed individual of the species called *Sepia officinalis*. The "bone" is enclosed within a membranous sac within the body of the cuttle, by which sac it is secreted, and with which it has no other connexion, dropping out when the animal is opened. On taking one of these objects into the hand, its extreme lightness is very evident, and if it be cut across and examined through a lens, the cause of the lightness will be perceived. The plate is not solid, but is formed of a succession of excessively thin laminæ or floors of chalk, each connected with each by myriads of the tiniest imaginable chalky pillars. When the cuttle is living, this structure runs through the entire length of the abdomen, being of equal length with it, and occupying about one-third of its breadth. In the Calamary the analogue of this object is of a horny consistence, semi-transparent, something resembling in shape the head of a spear, or the feather of a large pen, from which latter resemblance it is sometimes called the Sea-pen.

The well-known colour, sepia, is or ought to be manufactured from a black liquid, which is possessed by most of these creatures, and which can be ejected at will, probably with the view of darkening the water, and so temporarily baffling their enemies, of whom they have many, even including their own species. And it can also be employed while the animal is out of the water, as was once rather amusingly exemplified.

There was an officer employed, as I hope my readers have been and often will be occupied, in searching the coast for objects of marine natural history. After a while he came unexpectedly on a cuttle, who had taken up his abode in a convenient recess. The cuttle has a pair of very prominent eyes, and for a short time the cuttle looked at the officer, and the officer at the cuttle. Presently, the cuttle became uneasy, and taking a good aim at his military visitor, shot his charge of black ink with so true a range, that a pair of snowy white trowsers were covered with the sable fluid, and rendered entirely unpresentable. Even in many of the fossil cuttles this ink has been discovered dry and hard in its proper place within the creature. This most ancient substance has been removed, and ground down like very hard paint, and has been found to produce so beautiful a sepia tint, that an artist to whom it was shown inquired the name of the colourman who prepared it. And, in order to prove the character of the colour, a drawing of the fossil animal was made, and a description of it written, with its own ink.

Some of the Cephalopods are gifted with great powers of locomotion, and of those so gifted the Flying Squid is a good example. One of these creatures has been known to spring from the sea clear over the bulwarks of a ship and to fall on the deck, where it was captured. This specimen was six inches in length, and its habitation was the Pacific Ocean, lat. 34 N.

The eye of the cuttle is a most singular organ, its anatomy having long perplexed the dissectors, who could not conceive by what means the creature could see at all. It would be impossible here to describe this beautiful structure, and therefore I will content myself with observing that the celebrated Coddington Lens is merely a reproduction, though unwittingly so, of the lens belonging to the cuttle's eye. So here we have a pair of achromatic lenses and a series of air-pumps, contained in the structure of one creature belonging to the lower orders of the animal kingdom. Many other analogies exist, but space suffices not for them here.

The eggs of many fish are small and globular, being generally known under the name of spawn. The "hard-roe" of a herring furnishes a good type of this class of eggs. These, however, are the eggs of fish that are destined to be produced in countless myriads, and to serve as food for the other inhabitants of the deep, as well as for man. Uncounted thousands of these eggs perish before their maturity, being devoured by other fish which watch for them; and even when the young fry are born, comparatively few of them escape destruction. In order to compensate for such a loss of animal, the number of eggs is proportionably increased; one single cod-fish having been known to cast forth into the waves, in one single season, nine millions of eggs—equalling eight times the population of London.

THE COD-FISH.

But the destroying fish are not multiplied to such an extent, or the ocean would for a season teem with battles, and after a time be utterly depopulated. The eggs of such a creature as a shark, for example, are singly committed to the ocean; and in order to prevent them from being carried about at the mercy of the waves, or thrown to perish on the shore, they are of a most singular form. An egg of one of our British sharks, called the Dog-fish, is represented on plate H, fig. 4.

The egg is of a softish, horn-like consistency, so that it is not liable to be broken, or easily to be penetrated. The general shape of the egg has been aptly compared to a pillow-case, with strings tied to the corners; the enclosed pillow being the young shark. The long, curling, tendrilous appendages speedily affix themselves to sea-weeds, or other appropriate substances, and from their form and consistence anchor the egg firmly. In

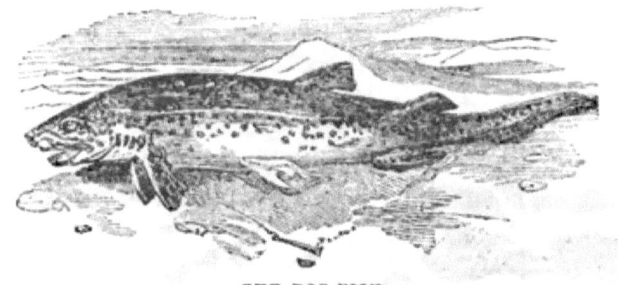

THE DOG-FISH.

order to enable the little shark to breathe, there is an aperture at each end of the shell, through which the water passes in sufficient quantity to renovate the blood. And in order to permit the enclosed fish to make its escape when sufficiently developed, the end of the egg nearest to the shark's head is formed so as to open by the slightest pressure from within. After the newly-born shark has left the egg-shell, there is no perceptible external change in its shape, for the sides are elastic, and immediately close up as before.

These eggs are popularly known by the poetical title of Mermaid's Purses—alas, empty!

There is another and more common Mermaid's Purse, that is found of various sizes and of various tints, although it is usually of a very dark brown, closely approximating to black, and in length from three to five inches. It will be at once recognised by the figure on plate II, fig. 1.

This egg is the production of one of the skates, and harmonises well with the strange, weird-like aspect of the creature from which it was produced. The accompanying figure represents the Thornback Skate (*Raia clavata*), a species that often attains a very great size, measuring some ten or eleven feet in length, and very nearly as much in breadth. From the four short arms

SKATE.

with which these eggs are furnished, they are thought to bear some resemblance to hand-barrows; and if a fisherman is asked the name of the object, he will generally call it "Skate-barrer." If the egg is picked up in the early part of the year, it will usually be found to contain the young animal—not a very prepossessing creature—as may be seen by reference to the engraving, where a portion of the egg is represented as removed. Perhaps the reader may remember Hogarth's "Gate of Calais," where a fisherwoman has upon her knees a huge skate, into whose countenance the painter has wickedly infused an expression precisely like that of the weather-beaten, withered old dame who holds it.

I was once talking about these eggs to some fishermen, who told me that in the spring they often found these eggs before the young were hatched, and were accustomed to boil and eat them just as hens' eggs are eaten. Whether to believe them or not I could not make up my mind, for fishermen are wonderfully loose in their details. However, as they gave me the information, I present it to the reader, and leave it to his own discretion to judge, or haply to his own energy to prove or disprove by actual experiment. I trust the latter.

In the summer months the eggs are invariably empty, or only filled with sand, little pebbles, and other shore *débris;* so that, unless the experiment can be tried in the earlier portions of the year, it cannot be made at all.

CHAPTER VI.

SEA ANEMONES AND OTHER ZOOPHYTES.

There is a singular class of animals, called by the scientific name of *Anthozoa*, or living-flowers, because their formation and external appearance seem to partake much of the vegetative nature. Among the most conspicuous of the Anthozoa, are the creatures called Sea-Anemones, although they are not in the least like anemones, but do bear a very decided resemblance to China-asters, daisies, or dahlias. Very many species of these curious beings may be found on various portions of the British coasts, especially those which lie quite to the south; some of these creatures are only to be found in certain localities, but there are two species which are to be found on every coast where there is the best shelter. These two species I shall describe.

On plate E, figs. 4 and 5, is represented the commonest of the Sea-Anemones, or Actinias, as they ought to be called; one figure showing the animal as it appears when covered with water and all its tentacles expanded in search of prey; and the other, as it appears when closed, and at rest. I regret to say that it has an exceedingly long title, and one which takes some time to say as well as to write, "*Actinia mesembryanthemum*," and which is generally curtailed in conversation or notes to "Mes," just as Mephistopheles was contracted to Mesty, by Mr. Easy, junior. Its popular name is the Smooth Anemone, it being so called from the smooth and slippery surface of its skin. When I say the "popular" name, I must be understood to mean the name popular among naturalists; for as to the people who live in close vicinity to it, they have no name for it at all. Very often, the fisherman has never seen such a thing as an actinia, or if he has been sufficiently observant to note it, he has no name by which he could describe it to another fisherman. Now, indeed, that so many enthusiasts crowd the sea-shores in their search after these actiniæ, and press fishermen into their service, a little knowledge on the subject is being gradually diffused; but some six or seven years ago, no one troubled himself about creatures which he did not catch with a net or line, and which he could not take to market.

The Smooth Anemone is generally the first that meets an observer's eye. As he wanders about among the rocks and stones that are left dry by the receding tide, he will see on many of them certain little lumps of green or red jelly, varying in size from a pea to a plum. On touching them, they do not feel soft, like jelly, but are very smooth, slippery, and much firmer in consistence than would be imagined from their aspect. In this state, some of them are entirely green or red, but the greater part are marked longitudinally with lines of different colours according to the variety. For there is but one species of "Mes," although the variety of colour pass through several tints of green and red.

If they are to be examined, they can be detached without injury, by slipping the thumb-nail or an ivory paper-knife under the base, and so gradually peeling them away from the support. Great care must be taken of the sucking-base by which the animal affixes itself, for a wounded base often causes death; and if the anemone has any nerves at all, as is but natural to suppose, their situation is on the base. Still, the process of peeling occupies some time, and if the tide is coming in, every minute is of importance. In this case, they can be removed instantaneously, by a judiciously aimed blow with the sharp end of the geological hammer, so that the portion of rock on which the animal is fixed is detached by the stroke. In this way a cargo may be collected in a very short time. They are generally found in colonies, consisting of five or six up to twenty or thirty in number; but now and then a giant may be found by himself or herself—for with them it is all the same thing—living in solitary state.

The smooth anemone seems to lead almost an amphibious life, for numbers of them are found affixed to stones and rocks in such situations, that they spend rather more time out of the water than in it. So that the rock is tolerably shady, the creature seems to be perfectly satisfied; but there are some hardy individuals who expose themselves to the full blaze of the sun. Even when they cannot be seen, owing to failing light, or rising tide, the touch will easily detect them, although in the latter case there is a slight chance of getting the finger nipped by a bad-tempered crab. I once filled a small basket with these creatures in a quarter of an hour or so, although the tide was driving me gradually to the shore, and there was not sufficient light to distinguish the hands of a watch.

The chief beauties of this, or indeed of any other of its relatives, cannot be seen until it is removed from the sea, and placed in a vessel where it can be subjected to light and exposed to close examination. While it remains in its native rock-basin, or adheres to its opaque support, the light cannot shine through the creature, and its base, one of the most remarkable features of its structure, is concealed by the stone on which it rests. If, then, the observer wishes to examine the anemone more closely, he may easily do so by acting as follows:—

Let him procure any glass jar—a large one is not at all requisite—and having chipped off a morsel of stone on which is growing a frond of ulva or porphyra, place it conveniently in the jar, and fill up with sea-water. After an hour or two, the sea-weed will be studded with tiny air-bells; and when that is the case the water is fit for the reception of the anemones, of which there should be only one or two. If the anemones are merely dropped into the water, they will sink to the bottom, and after a little while begin to crawl up the sides of the jar. Now is the time for examination, and with a lens of moderate power much of the structure may be made out.

Perhaps the "Mes" is the least attractive of all the anemones, but yet it possesses some extremely beautiful features which are peculiar to itself. When the creature is fully expanded, a row of little globules will be seen round the edge, among the tentacles. They are usually about the size of a No. 5 shot, and are blue and bright as turquoise, to which jewel they bear some resemblance. Indeed, if there is any distinction between the two, the animal turquoise is more beautiful than the mineral.

If a dissection is contemplated (and it will well repay the trouble), the creature can be killed by placing it in fresh water. Its anatomical structure is very remarkable, and even the muscular structure, that enables the creature to expand or contract at pleasure, is worthy of observation on its own account, independently of the fact that the presence of muscle is a proof that there must be nerves by which the muscles are excited. Although the details are complicated enough, yet the general notion of the creature is sufficiently simple, and may be readily imitated. Let a linen bag be made, and the mouth of it sewn up. Then let the closed mouth be pushed inwards, until the bag assumes the form of a cup with double walls, and it is the type of an actinia; the outer wall being the exterior of the animal, and the inner its stomach. The tentacles are hollow, and communicate with the space between the walls. This space is always filled with water, and by the contraction of the walls water is driven into the tentacles, and so expands them.

This is a capital species for an aquarium, as it will bear travelling, and is very hardy, enduring extremes of heat and cold bravely, but perishing immediately in pure fresh water. A rather remarkable circumstance connected with these creatures occurred some little time ago. A gentleman had brought some of them to town with him, and had been examining them in company with a friend. After the examination supper was brought by an unsophisticated servant, and removed by the same individual. While the table was being cleared the servant asked what was to be done with the anemones, and was told to put them carefully away in a jug. Now the only jug at that time on the table was a jug containing porter, and into that jug the anemones were severally dropped. About a fortnight afterwards, the anemones were again called into requisition, and the jug demanded. Great was the astonishment of their owner to see the porter-jug produced, and still greater when he found the creatures were still living. They have also been known to live in soapsuds for a considerable time.

If specimens are gathered for an aquarium, they should be chosen of a moderate size, for the larger actiniæ require a considerable bulk of water. They are in general very migratory in their character, marching at their own sweet wills over the sides of the vessel in which they are confined, and now and then paying a visit to the rocks, stones, and shells in its centre. Sometimes they will attach themselves to a frond of ulva, whose broad leaf affords a good hold for their base; and there they will stay for weeks. Sometimes they will not take the trouble of crawling down the sides, and then over the bottom, but turn themselves into boats by hollowing the base, and thus floating with the base upwards, until they think fit to contract themselves, and to sink. They generally remain for some time partially supported by one edge of the base against the side of the vase, but after a while they commit themselves freely to the water. Fresh-water snails may be seen floating in a similar manner.

The base of the actinia, by which it moves, and of which it is so careful, is a very pretty object when seen pressed against the glass side of a tank, and the light shining through its substance, showing the dark-green lines that radiate from the centre, and are so well contrasted with the azure gems that surround the disc. The mode by which the anemone travels is simple enough; it pushes forward one portion of the base, and, having fixed it firmly, draws the remaining portion after it.

There is a delicate gelatinous membrane that covers the entire animal, and which it frequently throws off. After an actinia has been sojourning in one spot for some time, and then moves away, it generally leaves a cast coat behind, as if to mark the exact locality of its habitation. Sometimes the creature appears to find a difficulty in getting rid of this membrane, which generally adheres strongly to the mouth. In such a case it is useful to assist nature, and a camel's-hair brush will generally give great aid to the actinia; who, in gratitude for help, expands itself immediately on being freed. Numbers of their cast membranes will be soon found in the aquarium, and should be removed.

If these creatures are kept merely for their beauty, they should be treated as greyhounds are treated; that is, kept almost entirely without food. They will live very well for many months without requiring food, but when it comes within their reach, they can devour and digest unlimited quantities. Perhaps the best plan is to give them a very tiny portion, which they eat, and which only stimulates them to protrude their hungry arms in the hope of getting more. Whenever I fed my own specimens, I generally gave them small pieces of beef, which they swallowed, and after a few days rejected, having extracted all the nourishment, and only left a white fibrous mass.

They were also very fond of flies, and ate many of them. Once a great bluebottle fly came buzzing into the room, and made such a disturbance that I immolated him, and gave him to the largest and hungriest anemone. About three or four days afterwards I saw the bluebottle floating on the surface of the water. On lifting it out of the tank it fell to pieces, being in fact the mere shell of the fly; all the interior having been, by some mysterious process, extracted in the stomach of the anemone. It is amusing enough to see the way in which an actinia eats. If a fly or a piece of meat be presented to its tentacles, it is instantly seized by them, and drawn to the mouth, the tentacle closing upon it on all sides. The animal then literally tucks in the morsel, and with it all the tentacles and the upper portions of its body, until they reach the stomach. Digestion then goes on very quietly, the presence of its own arms in its stomach being of no consequence at all to the animal; and in due time it untucks itself, and tosses away the indigestible portions of its food.

The "mes" multiplies readily in a tank. At the beginning of last September I had fifteen specimens in the aquarium, and by February in the present year there were between forty and fifty, many of them very minute, and very transparent. Indeed, scarcely a day passed without the discovery of a nursery full of little pink or green actiniæ, contained in an empty shell, or studding the surface of a shady pebble. Funny little things they are, and have a most consequential look as they spread out their tiny tentacles in search of food.

When the "mes" is caught for the purpose of stocking an aquarium, it will travel well if wrapped in wet sea-weed. If care is taken, the ulvæ, or enteromorphæ, that are used for this purpose, can be transplanted into the tank, and so two objects secured at once. The best plan is to put some wet green algæ at the bottom of a basket, and then to lay on them the anemones, each wrapped in wet sea-grass; over them another layer of green algæ should be placed, and so will they be quite comfortable.

The second species of Sea Anemone which I shall describe is, in my opinion, the most magnificent of the family, whether size or colour be the criterion. It is possessed of a scientific name hardly inferior in length to that of the smooth anemone, being called by the learned, *Bunodes crassicornis*. It is much too long a title for every-day use, and so it is contracted into " Crass." A portrait of a half-expanded crass may be seen on plate E, fig. 9.

This creature does not bear exposure to air and heat so well as the smooth anemone, and must be sought for in the shallows at low water. Sometimes at spring-tide a solitary specimen is seen, high, dry, and discontented; but such is an exception; usually it may be found just beyond ordinary low water-mark, expanding its gorgeous tentacles, and waiting for the numerous crustaceans or fish that are always left in the tide-pools.

The colours of this animal are very varied, hardly any two specimens being found of exactly the same tint, and the magnitude of its fully expanded disc is nearly equal to that of the crown of an ordinary hat. Some of them are scarlet, some pink, some lilac, some delicate grey, some of an olive-green, and all so delicately transparent that no colour can faithfully represent their beauty. On referring to the engraving, the reader will see that the base of each tentacle is surrounded with a pear-shaped, dark line. It is on this line that the depth of colour is chiefly lavished, the tentacle itself being always much fainter in tint, and as transparent as if formed from gelatine. The tentacles are very thick in proportion to their length, and it is from that peculiarity that the creature derives its name of "crassicornis," signifying "thick-horned."

The voracity of these animals is quite surprising.

I have often amused myself by watching them in their native haunts, and experimenting upon their powers of digestion. One single "crass," measuring barely three inches in diameter, required two crabs, each the size of a penny-piece, and a large limpet, before it ceased to beg with extended arms.

It is evident by the fact of the crab-eating that the crass must possess great powers of grasp, or it could never hold, retain, drag to its mouth, and finally devour, a creature of such strength as a crab of the size above-mentioned. Such a crab struggles with great violence, and requires a very firm grasp of the human hand to prevent it from making its escape. And yet the anemone, whose entire body is not larger than the closed hand, and whose substance is quite soft, can seize and retain the crab, if it is unfortunate enough even to thrust one of its legs within reach of the tentacles. There must, therefore, be some strange power by which this object is achieved; and the mode by which it is accomplished I now proceed to describe.

It has been long known that many water-inhabiting animals, of a class so low as to be scarcely more than water animated, can throw out long fishing-lines, of a substance so delicate as only to be descried by the sparkle of light upon them as they float about in the water, and that when these delicate lines even touch a little fish or crustacean, some power destroys that fish as effectually as if it had been struck by lightning. But this is not the mode of attack with the anemone, which resorts to means purely mechanical.

If the finger is brought into contact with the outspread tentacles of a healthy anemone, it will adhere to them with a peculiar, almost indescribable sensation. It is not in the least the adhesion of gum, glue, or any such substance, but bears some resemblance to that which is felt when the finger is thrust into the mouth of the common ringed snake, or of several fish. The mode of adhesion is only to be discovered by the aid of a tolerably powerful microscope, and when discovered is found to be equal in beauty to any structure that is yet known.

Scattered at intervals over various portions of the body, and especially crowded on the tentacles, are found tiny organs that are called by the name of thread-capsules. These are little oval vesicles, imbedded in the substance of the anemone, and containing within them a long delicate thread, closely coiled, in forms varying according to the description of capsule. The extreme tenacity of the thread may be imagined from the fact that the largest capsules are not more than the three-hundredth of an inch in length, and that within so small a compass the thread is coiled like a watch-spring in the barrel. Indeed, the simile of a watch-spring will nearly express the object, for the thread is so strong, in spite of its tenuity, that it has aptly been compared to the hair-spring of a watch. When the tentacles are irritated or compressed, myriads of these capsules start forward, become everted, and shoot forth their tiny spears. The length and shape of these wonderful filaments are very various, some being of a very great length, and so fine that a microscope of high power can hardly distinguish them; while others are only two or three times the length of the capsule that contained them, and covered with an armature of short hairs even more minute than themselves.

It is easy enough to see these singular organs, even with the aid of a good hand-lens; but with a microscope nothing more is required than to cut off part of a tentacle, get it well in the field of the microscope, and then apply pressure: some of the capsules will dart forth their threads almost immediately, but others will require greater force before they will evert themselves.

The crass may often be passed by without observation if it does not choose to display itself; for when closed nothing appears but a round mass of sand and broken shells, about the size of a penny-piece, and projecting so slightly above the sand that an inexperienced eye would see nothing remarkable in it. This unpromising aspect is assumed when the creature has had a large dinner, or when it is alarmed. It is rather curious to see how suddenly a magnificent specimen of living flower collapses into a shapeless, and not at all pleasing knot of sand, stone, and shell. The crass is a delicate animal to preserve, for it seems to require a large body of pure water for respiration, and if in the least injured does not recover from the wound like the smooth anemone. If it were as easily detached as the "mes" there would be less difficulty in preserving it; but it has an unpleasant habit of forming its base into some six or seven lobes, attaching four or five of them to separate stones or portions of rock, and pushing the others into any crevices that may be convenient.

Very seldom, indeed, are the fingers alone sufficient to extricate the creature without injuring the base, and unless that important part be preserved in its integrity the crass assumes various wonderful shapes, and very soon dies. In such a case it generally begins by puffing out several striped lobes from its mouth, which project, and soon assume so enormous a size that they quite overshadow the tentacles, and render it a matter of some small difficulty to ascertain clearly whether the animal is not reversed. After it has thus displayed its own temper, it proceeds to try that of its possessor, by assuming various forms with such rapidity, that to draw it with any accuracy is quite out of the question. In the morning it may present a splendid and regular disc of tentacles, expanded to their utmost, and glowing with scarlet, pink, or lilac: just as a rough sketch of the creature is taken, it half withdraws the tentacles, and begins to puff out the striped lobes or lips. Another sketch is now taken of the crass, as it appears when in a pouting humour; but there is no time for any colouring, because the perverse animal now begins to take in water to an alarming extent, and soon succeeds in turning itself into an animated hour-glass, hardly a vestige of tentacle or pouting lips being visible. Even this form does not seem to please it, and is speedily exchanged for another, which in its turn gives place to a fifth, and so on *ad infinitum*.

In order to avoid the danger of tearing the very sensitive base, I was in the habit of digging out all the stones and pieces of rock to which the crass had affixed itself, and permitting the creature to free itself in the tank; a feat which it generally soon accomplished. By taking these precautions I succeeded in bringing a very large and gorgeous specimen to town, and preserved it alive and healthy for upwards of six weeks, in one of the most crowded parts of the city. It would probably have lived much longer, had not the density of the water in which it resided been corrected too suddenly. The animal was in a large vase of seawater, which necessarily presented a wide surface to the air; and so the water rapidly evaporated, leaving all its salts behind, and rendering the remaining water much too dense. In the correction of this evil the fresh water was added too rapidly, and inflicted on the nervous system of my poor crass a shock from which it never recovered.

In such a case the tentacles begin to droop, they then lose their beautiful transparent colouring, and become dull, opaque, flaccid, and exceedingly small. Whenever the creature is in this state it is not easily restored to health and brilliancy.

This species does not travel in wet sea-weed nearly so well as the smooth anemone, and besides, pours out a vast amount of mucus, which makes the whole of its neighbourhood exceedingly unpleasant. In order to transport it in perfection it should be indulged with a jar of sea-water, and have no travelling companions. Its tentacles are so strong, and the animal is so voracious, that it will frequently destroy any other creature that happens to be placed in the same vessel. The very specimen which I have described, killed, during the few hours of its journey, a beautiful Æsop prawn, on which I rather prided myself, and a young gurnard. It did not eat either of these creatures, but simply caught them, and retained them until they were dead. How they were killed I do not exactly know; but it is suggested with some reason, that the capsular threads convey with them a kind of poison, that is effectual enough among the smaller animals, but not sufficiently powerful to affect human beings.

The coating of little stones and broken shell has already been mentioned. These substances are evidently chosen instinctively with a view to concealment, and are fastened to the body by little sucker-like protuberances, with which the greater part of its surface is studded. If the animal is used for culinary purposes, for which it seems to be adapted as well as the oyster or the periwinkle, this shelly coating must be removed; an operation which is easily enough effected by the fingers, and not so tedious as plucking a fowl. There is no difficulty in finding a sufficient number to form a respectable dish, for any one who knows where to search, and what to see, may capture an unlimited supply in an hour or so.

I conclude this short history of the creature by observing that the reader must not expect to find the crass presenting precisely the appearance of the specimen here depicted, for the whole of the body is generally invisible when the creature is offended, part being buried in the sand, and the remainder overshadowed by the thick and close-set tentacles. If, however, it is carefully removed, and takes up its residence in a jar or tank where there is no sand, it then assumes the shape of the figure, at all events for a time; but after a little experience of prison it changes its shape so frequently that it almost realizes the fable of Proteus, and his efforts in a similar difficulty.

On plate L, fig. 1, is a representation of another British zoophyte, allied to the anemones, but yet distinct from them. It is the Common Madrepore (*Caryophyllia Smithii*), a member of that wondrous family that produces the coral and similar substances.

When removed from the water, or even when alarmed, the animal portion of the madrepore suddenly shrinks, and little is visible save a series of thin calcareous plates, standing on edge, and radiating from a centre. But when the creature has recovered its self-possession, and begins to feel rather hungry, a beautiful semi-transparent living substance emerges from the chalky plates, and after a while puts forth a number of tentacles, tinted with most delicate hues, and much resembling those of the anemones, except that each tentacle is terminated by a little globular head. These tentacles, like those of the anemones, are covered with filiferous capsules, and adhere to the hand in much the same manner, though not so strongly. This madrepore is voracious enough in its own way, but does not seem to care very much for food supplied artificially. I had a specimen alive for some months, but could not get it to eat any but the minutest portions of meat, while it would have nothing to do with a small fly; yet it was healthy, and almost constantly protruded its transparent tentacles. The proximate cause of its death appeared to be attributable to a bad-tempered Daisy Anemone (*Actinia bellis*), which lived in a cave like a hermit, and did not approve of intrusion. I am accustomed to stir the water daily, in order to imitate as far as possible the natural stir of the sea, and in so doing this madrepore was washed into the cave where resided the daisy anemone. Although it was speedily replaced in its own little home, which had been specially chiselled in the rock, it never properly recovered; and after leading a dull, inactive, colourless existence for a week or two, fairly died, and left me nothing but its skeleton as a memorial.

The madrepore may be found adhering to, and in fact almost forming part of, the rocks, requiring the aid of a strong knife to detach it without injury. The specimen represented in the engraving is rather larger than the general run, and only exhibits the tips of the tentacles, and the expanded membrane that edges the calcareous plates. When the creature is dead, or alarmed, the arrangement of these plates is plainly visible, reminding the bystanders of a reversed mushroom suddenly petrified.

There is a certain substance often found on the shore, tough, soft, fleshy, and unprepossessing, well deserving the popular name that is given to it, namely "dead man's fingers," or "dead man's toes," as the case may be. For, just as the smooth anemone when touched collapses into a shapeless lump of green jelly; as the "crass" shrinks into the sand, an almost undistinguishable and unrecognisable excrescence; as the madrepore retires within its own chalky walls, so does this zoophyte withdraw all its beauties in the uncongenial conditions of heat and drought, and only present an exterior anything but agreeable to sight or touch. The scientific name of this zoophyte is *Alcyonium digitatum*, or the Finger-shaped Alcyonium, and a representation of it may be found in plate E, fig. 8, accompanied by a magnified sketch of one polyp.

In this and other zoophytes, the qualities, fraternity and equality, are exhibited in a manner far superior to any republic, ancient or modern; but there is very little liberty in the case. In these curious creatures communism prevails to its fullest extent, one for all and all for one. There is one body, so to speak, the polypidom as it is called, and from this body protrude, under favourable circumstances, innumerable polyps, each one gathering its nutriment from the surrounding water, and conveying that nutriment not to its own body only, and for its own aggrandisement, but into the general polypidom, affording to each of its thousand relatives a portion of its own nourishment, and receiving from each of them some modicum of their own.

When placed in clear sea-water, the alcyonium soon begins to put forth a few crystalline columnar polyps, each standing boldly out, and bearing a mouth or head, composed of eight radiating, slender, pointed petals, fringed with delicate hairs. The internal anatomy of this creature, or rather of this creature-mass, is very interesting, and worth studying practically.

On plate E, fig. 1, is shown an example of a plant-like, compound animal, very common on the coasts, either thrown on the shore dead or dying, or affixed to large algæ, near low water-mark. This is the Sertularia, a beautiful family of zoophytes, of which some sixteen or seventeen species are found on our own coasts. The species represented is perhaps the commonest of all, *Sertularia filicula*.

If one of these creatures is examined by the aid of a moderately powerful lens, it will be seen to consist of a horny, many-branched stem, each branch being studded with a double row of little cells, open at the mouth, which is much smaller than the base. If the creature is placed in clear sea-water, and still watched through the lens, each cell will be seen to protrude a tiny polyp, whose star-like head is all that is visible externally. The polyps are easily alarmed, and in such a case withdraw themselves wholly within their cells. There is a very similar zoophyte called by the name of *Plumularia*, which may, however, be easily distinguished from Sertularia, by the position of the polyp cells, which only occupy one side of the branches; whereas those of Sertularia are to be found on each side equally, sometimes in pairs, sometimes alternately, according to the species. When dried, both these creatures retain their chief characteristics, and if any of the sea-weed landscapes be examined, both Plumularia and Sertularia will generally be found among the algæ. They are delicate in constitution, and not easy to keep in an aquarium, unless under singularly favourable circumstances.

There is another pretty, plant-resembling zoophyte, found plentifully enough near low water-mark, but at a rather higher elevation than the preceding. This is known among zoologists as *Coryne pusilla*, and is chiefly remarkable on account of the peculiarity from which it derives its name. "Coryne" is a Greek word, signifying a club, or knobbed stick, or more properly a mace, such as the steel-headed, spike-armed weapons with which our gentle ancestors were in the habit of exciting the brains of their adversaries.

The stalk of this zoophyte is about as thick as ordinary sewing thread, and it clings to the sea-weeds among which it resides as much as the cotton in question would do. Even when seen with the unassisted eye it is rather an elegant creature, but when a lens is brought to bear upon it, sundry hidden beauties become obvious, and among them that peculiar formation of the polyp from which it has derived its name of Coryne. Its club-shaped head is studded with numerous tentacles, that are arranged in a manner somewhat similar to the steel spikes of the war mace. Each of the tentacles is furnished with a globular head, and if submitted to a higher power of the microscope, appears covered with minute knobs, at the extremity of each of which is a short straight bristle. The stalk is merely a horny tube, ringed in structure, and increasing in size towards each polyp head, so as to allow room for them to change their position. The movements of these creatures are not very rapid, but can be clearly seen. A magnified representation of a single polyp head accompanies the figure of the zoophyte.

We now come to one of the most remarkable objects in the whole range of animated creation, and which requires a microscope of some power to develop. On plate E, fig. 7, may be seen a kind of miniature tree; this is the representation of an elegant little zoophyte that is found plentifully near low water-mark. It is small, seldom exceeding two or three inches in height, and is of a delicate and feathery texture. To the naked eye there is nothing of any great importance in this creature, but if a portion of a branch be detached and brought under the field of a microscope, a very strange sight meets the eye. As is the case in all these zoophytes, the branches are studded with cells, in which live little polyps, at one time expanding wide their feathery tentacles, and at others sulkily gathering them up like an unsuccessful fisherman gathering his net on his arm. But to each of the cells is attached a most singular appendage, precisely resembling the head of a bird and one joint of its neck, by which it is attached as a point to the cell, and on which it works. There is certainly no eye in the head, but there is a most decided beak, which opens and shuts precisely like the beak of a bird, while the entire head keeps up a continual nodding backwards and forwards on its joint. If the portion selected is tolerably healthy, it will contain from ten to fifteen, or even more, of these birds'-heads, bowing to each other in the most oppressively polite style, and every now and then shutting their beaks with a sharp snap.

By the side of the zoophyte itself is shown one of the bird-head appendages, as it appears when fastened to the cell. The object of these strange organs is not at all clearly ascertained, for they seem to have but little connexion with the polyps that inhabit the cells, and bow with as much perseverance when the cell is empty as when it is occupied by its living inhabitant. Many zoophytes possess the bird's-head; but as that species which I have mentioned is perhaps the most common, and is easily detected, it has been admitted as the representative specimen. It must be understood that when these creatures are subjected to the action of a microscope, they must be well supplied with water, or they will die speedily. If the power is not very high, the entire creature may be placed in a flat glass cell, and slightly compressed against the side by a glass plate. The microscope should then be set horizontally, and will show the structures tolerably well. But if a higher power is requisite, a portion must be removed, and placed in the animalcule cage with which every good microscope is furnished. A flat watch-glass, and a piece of the thin microscopical glass used for covering preparations, will make a good extemporised animalcule cage, if the regular machine is not at hand.

Among the other members of the animal kingdom that are popularly ranked as vegetables are the *Flustræ*, for an example of which see plate E, fig. 2. These creatures do, indeed, much resemble the leaf of some plant, and so closely that uneducated people can hardly be made to believe their animal origin. If, however, the fingers are passed over the surface of the flustra, and especially if they are passed from the point of the leaf towards the base, a peculiar rough, harsh, and stony sensation will be perceived, and this sensation is caused by the innumerable spine-crowned cells with which the leaf is covered, and, indeed, of which it is composed. A close examination with the naked eye shows that there is a curious structure not usually met with in plants; but if an ordinary pocket-lens is brought to bear on it, the entire surface will be seen to be composed of little oval cells, arranged in rows something like the scales of a fish, or tiles on a house-top. Each cell is armed with four short, sharp spines, that project from the upper portion of the edge, two at each side, and these spines are the cause of the peculiar rough sensation that is communicated to the finger. The cells are placed back to back, like those of a honeycomb, so that there is no right or wrong side to the flustra leaf; for leaf it must be called, although its proper title is polypidom. The species given in the plate is *Flustra foliacea*, a very common zoophyte, and often found on the shore, thrown up by the sea in large masses.

On many sea-weeds may be found a kind of stony scurf that spreads over their leaves or stems, and often destroys the beauty of the specimen. It is true that the appearance of the alga may be injured, but the creature that injures it is of so curious and beautiful a form that it ought to be preserved. This stony scurf is called Lepralia, and consists of innumerable cells, not unlike those of the flustra, spread evenly over the surface of the substance to which it adheres, and often so thickly that there is hardly a spot left uncovered. I have now a remarkably fine specimen of *Delesseria sanguinea*, measuring eighteen inches across, the whole of whose stem, and great part of the leaves, is overrun by several species of Lepralia, which, although they certainly rather disfigure the plant as a dried specimen, look so lovely under the microscope that I would not on any account have them away.

A common species of Lepralia is depicted on plate E, fig. 6. The upper figure represents the zoophyte as it appears when magnified about thirty diameters, equalling nine hundred times superficially; and the lower figure shows its appearance when slightly magnified by a simple pocket-lens. Almost every sea-weed of any dimensions, and especially the *Laminariæ*, will be partially covered with the cells of this zoophyte, and it may also be found on shells and other objects which have been submerged in the sea for any length of time. There are nearly forty British species of this single genus, and if half a dozen specimens are examined, it will probably happen that three or four will be distinct species.

CHAPTER VII.

STAR-FISHES AND SEA-URCHINS.

PEOPLE seem to have a strange love for comprehending various descriptions of objects, whether animal, vegetable, or mineral, under a single term, and generally contrive to hit upon a word which could not be rightly applied to any of them. Take, for example, the word "fish," as we are speaking of marine objects. Not to mention the whale, and other cetaceans, which are popularly called by the name of fish, we have lobsters, crabs, shrimps, oysters, limpets, mussels, &c., all comprised in the term "shell-fish." Then, we talk of cray-fish, cuttle-fish, jelly-fish, and star-fish, not one of the whole party having the very smallest right to the title of fish. Still, custom has so inextricably woven the name into the idea, that they cannot well be separated, and therefore must be retained until the same power shall unweave its own web. These prefatorial remarks must be my excuse for employing the word "star-fish" in the present case, and "jelly-fish" in a succeeding chapter.

Every one has heard of star-fish, and most people have seen them, either in a preserved state, or as they appear when thrown up by the waves. There are very many British species of star-fishes, but out of them I have chosen three, as types, to which, indeed, most of the species can be referred. The commonest of the British star-fishes is the Five-finger (*Uraster rubens*); for a figure of which see plate I., fig. 4. There are few days when some of these creatures are not cast on the shore, and there left by the retiring tide, so that their habits and anatomy may be easily studied. Generally they appear to be dead, and, indeed, sometimes are so; but their apparent death is often but quiescence, and if they are placed in sea-water they become lively in a very short time.

If a star-fish is thus rescued, and laid in a shallow rock-pool, where its movements can be watched, it will give ample food for contemplation, were it only for the mode in which it moves from one place to another. This movement is very slow, gentle, and so regular, that the eye cannot detect any motive power at work. Should a stone, a ridge of rock, or any other impediment, be in the path of its progress, the star-fish does not seem to trouble itself in the least, but continues its still, gliding movement, as quietly as if it were moving on level ground. As the stone is reached, one ray of the star-fish is gently pushed upwards, and seems to adhere to the stone; another follows, then a third, and presently the creature is seen to climb the stone with quite as much ease as if it were walking on level sand. The rays accommodate themselves, in a very curious manner, to the shape of the substance over which the animal is crawling; so that if it is passing over a sandy spot, interspersed with furrows and pebbles, the arms of the star-fish never bridge over the furrows, but pass down one side and up the other, while precisely an opposite process takes place with regard to the pebbles. The star-fish can thus climb rocks that are perpendicular, and clings firmly even when they overhang. How this process is conducted we shall presently see.

I may as well remark here, that if the star-fish were dead when put into the water, the observer would find his patience well tried if he waited to watch for these movements, so that he ought to be quite sure whether he has hit upon a specimen that is living. Now, however dead a star-fish may appear to be, if it is of a tolerably firm consistence to the touch, it is a living being, even though there should be no perceptible movement. If, however, in taking it up, it hangs loose and limp, life has departed, and it can only be used as a specimen for preparation, or for anatomical purposes. Any doubt will soon be settled, by placing the dubiously vital animal in clear sea-water for a few minutes, and then suddenly turning it over. If there is any life remaining, the numerous feet that occupy the under surface will move about, and the creature will soon recover its wonted activity.

When a living star-fish is laid on its back, a number of semi-transparent, globular organs will be seen in constant movement, being thrust forward and then withdrawn, moving from side to side, as if feeling for something, as indeed they are. These are the ambulacral organs, as they are scientifically called, but I prefer to call them feet, on account of their office. These feet are, in fact, suckers, and can be protruded or withdrawn by a very curious piece of mechanism, which is not easily described without the use of diagrams, but which I will endeavour to explain, as far as possible, without them. The feet are hollow tubes, each passing into the interior of the animal through a circular aperture, and being furnished with a globular, membranous head, just within the skin, filled with a fluid; so that, in fact, each foot or sucker bears some resemblance to a brass-headed nail driven through the skin, the head remaining within the animal, and the nail itself projecting. Now, if the creature compresses the membranous head, the fluid contained within it, being comparatively incompressible, seeks an exit, and finds none except the hollow of the tube. Through this tube it accordingly runs, and so pushes forward the sucker which terminates the foot. When the pressure is removed, the fluid returns into the head, and the sucker is retracted.

The mouth of the star-fish is placed underneath, and in the very centre of the body, the stomach being immediately beyond the mouth, as is the case with the sea-anemones. The stomach does not seem to occupy very much space, but it is capable of accommodating a large amount of nutriment, in which it is assisted by certain supplementary stomachs, which run through each ray, nearly to its extremity. These supplementary stomachs, or cæca, as they are called, may be seen by slitting up the skin of the upper surface of the rays, when the cæca will be seen lying immediately beneath, looking more like dark, loose, unformed masses of liver, than mere appendages to the stomach. And it is a very remarkable fact, that although these large and very important organs exist in each ray, the star-fish appears to be indifferent to the loss of one or more rays, and fills up the wounded space so perfectly that it is hardly possible to distinguish the spot where the ray once was. This circumstance accounts for the fact that star-fishes, apparently perfect, but only possessing four rays, are sometimes met with, whereas the minimum of number is five. When the star-fish has been seen to cast away its rays in captivity, the amputated rays still continued to move their feet-suckers as when they were attached to the body, but they did not appear to be capable of continuing their march.

Small as the mouth of a star-fish appears to be, small as is its stomach, and feeble as are its muscular powers, it can swallow a bivalve mollusc entire, or if needful, open it, and suck out the contents in some mysterious way; a feat that no man could accomplish without tools. Even with the proper knife, oysters are not very easy to open without some practice; but if a man's food were restricted to oysters, which he must open without the assistance of any tool, he would run considerable risk of starvation. The ancient naturalists were well aware that the star-fish possessed the power of eating oysters, but they thought that the creature accomplished its design by watching until an oyster opened its shell, and then poking one of its rays between the shells as a wedge; then, having once gained a partial admission, it slowly insinuated itself, and finished by devouring the inhabitant. It appears, however, by the reports of careful observers, that the oyster-eating is true as to the fact, but false as to the mode. The star-fish seems to bring its mouth in contact with the edge of the shell, and then from some delicate vesicles never protruded at any other time, to pour into the oyster some drops of a poisonous fluid, which forces the animal to open the shells, and finally kills it. Such is the account as it stands at present.

The skeleton of the star-fish is one of the most complicated structures imaginable, much too complicated for description here. It may easily be obtained by any one who wishes to possess such an object, if he takes a perfect specimen of the creature, and places it near an ants' nest. In a very few days, the ants will nibble it to pieces with their sharp, sickle-like jaws, and eat away every particle of the soft portions, leaving only the skeleton, which will then look like a singularly beautiful specimen of carved ivory. Ants, by the way, are very useful insects to the naturalist, and are capital skeleton developers. Only they do *not* store up the food in their subterranean mansions, as is popularly imagined; for as they feed on animal substances, and not on corn, their stores would soon be exhaled in the form of gas. There are always plenty of ants' nests near the coast, and it would be useful to look out for them as soon as possible, taking care to choose those that are not exposed to the public gaze, or near a public path. I would recommend the use of a box, perforated with many holes, as a convenient mode of keeping the specimens from dust, and at the same time of permitting free access to the ants. I had, until lately, an exquisite skeleton of a lark that had been prepared in a similar manner.

The colour of the five-finger star is generally a dusky red on the upper surface, the colouring matter of which is sometimes irritating to those who possess delicate skins. Sometimes, however, specimens are found of a purple or violet hue, and are by some authors considered to be a distinct species, although they are probably but a variety of the common red five-finger. This species, especially if large, is not very suitable for an aquarium, and seldom survives for any length of time. One individual that I tried to domesticate had, for the last few days of its life, a curious habit of resting merely on the points of its rays, and elevating the disc in the centre, so that it presented somewhat the aspect of a five-legged table, the rays forming the legs and the disc the table itself.

There is another tolerably common species, that is found on the shores, and is very different from the Five-finger, being composed of a large disc, with twelve short pointed rays proceeding therefrom; so that when a large specimen is seen in scarlet splendour on a rock, it seems to blaze out like the sun, and has accordingly been called the Sun-star. It is not easily mistaken for any other creature, but in order to make its recognition easier, a figure of it will be found on plate L, fig. 5. Its scientific name is *Solaster papposa*. Its usual colour is a bright red, but it is often seen to be tinged with violet, while in some specimens the rays are very much paler than the disc, and one most singular example has been recorded of a mixture of bright green.

Occasionally near low water-mark may be found specimens of a very curious star-fish, differing as entirely from the sun-star, as it from the five-finger. This is the Brittle-star, of which there are several British species: the commonest of them, *Ophiocoma rosula*, is given on plate L, fig. 3, and as will be seen, is a very curious creature; its form has been well described by the image of a little sea-urchin, surrounded by five very lively centipedes. Indeed, it hardly resembles the sun-star at all; but these creatures assume such singular shapes, that forms the most dissimilar are found actually to be closely united;

this we shall see presently when we come to the urchin. The *Ophiocoma* is called the brittle-star on account of its inexplicable custom of breaking itself into little bits when alarmed. It is really a matter of some little difficulty to secure an entire specimen, so that a really perfect brittle-star is rather a valuable acquisition, though the creatures are so common that a dredge will haul them up by pailfuls. One of the largest British species of star-fish, *Luidia fragilissima*, a creature measuring some two feet across, possesses this suicidal property in a high degree. In Forbes's British Star-fish, a work which I strongly advise all naturalists to obtain, or at least to read, is a most ludicrous account of an adventure with a *Luidia* and a bucket.

When undisturbed in their own element, the brittle-stars are well worthy of observation; for their long fringed arms wriggle about with great vivacity, and well carry out the simile of the centipede. In order to destroy these creatures without damage, a vessel of fresh water should be brought to them, and if they are rapidly submerged, the saltless fluid destroys them before they have time to discover that there is anything wrong. In the case of the *Luidia*, however, the sight (if star-fishes see) of the fresh water was so alarming that the precautions were useless.

If the star-fishes are needed for the cabinet, they must be dried; an easy process enough, but requiring patience. They must first be thoroughly washed in fresh water, in order to get rid of the salt, and then carefully spread out on a clean smooth board, and dried in the open air. They should not be placed in the cabinet until they are thoroughly dry, or there will be sad damage done.

I said just now that we should soon find that forms, apparently dissimilar, were in reality closely connected with each other; and this fact we shall see exemplified in the creature that next comes before our notice, the Common Sea-egg or Sea-urchin (*Echinus sphœre*) whose external appearance is shown on plate L, fig. 2. There does seem to be some slight connexion between the three star-fishes which we have just examined; but that there should be any connexion at all, or any relationship, between the brittle-star and the sea-urchin, appears too preposterous an assertion for credibility: yet such is really the case, as will soon be seen. The specimen from which the drawing was made is a tolerably perfect one, being still furnished with its array of spines from which it derives its name of sea-urchin, the urchin being a popular name for the hedgehog. But if these spines are rubbed away, a smooth surface will be left, on which are numerous tubercles, marking the spots on which the spines formerly rested. If now the reader will take a damaged urchin, plenty of which are to be found on the shores, and examine its external appearance, he will see that it has a very close relationship indeed with the star-fishes. Let a common five-finger star-fish be laid on its back, and the points of its rays stitched round a little disc of leather, it will then assume very much the aspect of a skeleton urchin. Let then the spaces between the rays be filled up with a substance of the same structure as the rays themselves, and then we have a complete Echinus, complete, at all events, as to its general external appearance.

The specimens that are found cast up by the waves are generally destitute of the spiny armature that is found upon them in their living state, and thus permit the eye to perceive the formation of the shell. Close by the figure of the Echinus itself will be seen a little diagram composed of several angular forms studded with little tubercles. These represent the pentagonal plates of which the shell consists, and which are most wonderful instances of animal economy. In the shell of every Echinus are hundreds of these plates, varying in size according to their position, and so closely connected with each other that externally their marks of junction are not perceptible; but if the shell is broken, and examined from the interior, the shape of these plates becomes tolerably well defined. It will be observed also, that when the shell is broken, the serrated edges of the fractured portions show the angular form of the plates. As the shell is composed of these plates, it may well be asked how the creature can possibly increase in size, because it cannot, like the lobster and other crustaceans, throw off its old coat when too small, and take to itself a better; and to add to the difficulty, there is no supply of arteries and veins ramifying through these plates, as is the case with the bones of a vertebrate animal, but each plate is dense and dead.

In order to overcome these apparently insuperable difficulties, a very beautiful arrangement takes place. The delicate living membrane with which the entire surface of the body is covered insinuates itself between the edges of these plates, and continually deposits round the margin of each particles of calcareous matter; so that each plate simultaneously increases round its edge, and the original form of the shell is preserved.

If we still keep before our eyes the image of the rolled-up star-fish, we shall see that as the mouth is precisely in the centre of the disc, it would also be found in the centre of the Echinus shell. And such a mouth as it is could hardly be conceived. If a human being, say a man of six feet in height, were to be possessed of a similar mouth, it would be about the size, and very much the shape, of an ordinary wooden pail, the teeth being as long as the staves of the pail, only they must be made very sharp at the top, and but five in number. The teeth of the Echinus may be seen protruding from the mouth, and their extreme hardness may be tested by the finger without any danger. The entire arrangement of teeth and muscles, and bony scaffolding, is so exceedingly complicated, that even with the help of diagrams it would be difficult to explain the structure, and without their aid quite impossible. I may, however, mention, that there is some resemblance between the teeth of the Echinus and those of rodent quadrupeds, there being a provision for adding fresh substance to the tooth as fast as it is worn away by use.

If the reader will now examine the interior of an Echinus shell, he will see that it is marked out into five equal parts, by five double rows of perforated plates, containing many hundreds of very minute apertures. Through these apertures protrude sucker-feet, just like those of the star-fish, which have already been described, and worked in the same manner.

Reverting now from the interior to the exterior, we shall find its surface thickly studded with spines, which, as well as the suckers, are employed as a means of locomotion, and therefore must be freely movable. If a single spine be removed, and note taken of the part which it previously occupied, it will be seen that on the shell is placed a rounded tubercle, and that the base of the spine is furnished with a hollow socket into which the tubercle fits, so that the spine has perfect facility of movement. The spine is bound to the tubercle by a short tendinous ligament, connecting the centre of each, much as is the case with the larger joints of vertebrate animals. The power of motion is communicated by the membranous covering that envelopes the body during the life of the animal. The spines of some foreign species of Echinus are very delicate and sharp, piercing the unwary hand like so many needles.

Besides all these multiplied means of progression, there are other very tiny organs, which may possibly be used as assistants for the same purpose, or they may possibly perform some office at present unknown. Among the spines there may be seen, with the assistance of a lens, a very great number of little three-headed pincers, standing each upon a flexible footstalk. These are called *pedicellariæ*, and are also found on several of the star-fishes. Some naturalists have regarded them as distinct animals, residing parasitically upon the Echinus. The Echinus is often boiled and eaten, just like eggs;

from which circumstance it is sometimes called the Sea-egg. All these creatures are called by the general name of *Echinodermata*, signifying "urchin-skinned." We have already seen how great is the apparent distinction between the several creatures that are classed together under this title, and this outward distinction is quite as great in the last example that will be here mentioned.

There is a curiously-shaped creature represented on plate F, fig. 6, looking something like a cucumber, with a feathery fringe attached to one of its ends. This is popularly called, on account of its shape, the Sea-cucumber, and is scientifically termed *Holothuria*. The derivation of the word is Greek, but its signification is very uncertain. It was used by Aristotle in his "History of Animals;" but the reason why he so named the creature no one can tell.

The holothuriæ are very curious creatures, for they possess some organs of the Echinodermata, by virtue of which they rank as Echinoderms, and they also have other organs, which seem to imply a connexion with the sea-anemones, while the shape somewhat approximates to the annulate form of the worms. Like the star-fishes and the Echinus, they possess five rows of sucker-feet along the body, although in some species these sucker-feet are scattered over the entire surface. If a holothuria be opened, almost the entire cavity of the body is filled with small white tubes, which are apt to tumble out, and become inextricably confused, if care is not taken. Indeed, at first sight, a freshly-opened sea-cucumber reminds one of the famous cucumber in the "Arabian Nights," which was stuffed with pearls by command of the talking bird, only that the sea-cucumber appears to have been stuffed with white bobbin. These white threads are the egg tubes. Altogether, it does not present the most inviting aspect to the eye, nor does it appear to be a very suitable object for the table. Yet it is one of the favourite dishes of that omnivorous nation, the Chinese, who pay large sums for fine specimens.

When the holothuria feels unwell, or is displeased, it has a very remarkable habit of dispensing with its teeth, stomach, and entire digestive apparatus, and so converting itself into a mere empty bag, with an useless mouth at one end of it. However, animals of this order are not easily killed, and before very long a fresh set begins to grow, and in a few months the holothuria is as perfect as ever. The beautiful feathery plume that surrounds the head, or rather the mouth, is the organ that has caused some naturalists to class the creature with the actiniæ or anemones, to whose tentacles the plume bears so close a resemblance. In general, the body of these animals is too thick, and the skin too tough, for the adoption of the suicidal habits of the star-fishes; but in some of the species, whose diameter is very small in proportion to their length, the creature actually does succeed in breaking its body into several fragments. The reader may compare with this habit the similar custom that prevails among many lizards, of snapping their tails off if they are touched or suddenly alarmed; as is especially exemplified in the case of the Common Blind-worm, which is often known to break itself across, as if it had been made of glass. If any of these creatures are found in a living state, they will not at first put forth their tentacular crown, and the owner must be content to wait. But if they are properly supplied with clear and pure sea-water, they will generally exhibit a large portion of the tentacles, if not the whole. They may usually be found clinging firmly to stones and pieces of rock, in situations where they are not exposed to light; for the influence of light seems to be exceedingly painful to them.

CHAPTER VIII.

ANNELIDS—BARNACLES, AND JELLY-FISH.

On plate F may be seen some figures of strange-looking creatures, having a kind of general resemblance to each other, but belonging to very different ranks in the animal kingdom. They are, however, placed near each other, in order to show how the same idea of form runs through different genera. The figure on the left is no worm, although it bears some resemblance to the creatures whose portraits occupy the top and right of the same plate. No. 3 is a mollusc, ranking with the periwinkle, mussels, nudibranchs, and other creatures, which have already been described in Chap. II. Then, again, the figure occupying the bottom of the plate has rather a worm-like aspect, and would bear even a close likeness, if it were much longer in proportion to its diameter, as is the case with some of its congeners. This animal, however, belongs to the star-fishes. The central figure, which is really one of the worms, looks much more like a common garden slug than an earthworm; to which latter creature, however, it is in near relationship. One may therefore easily pardon the errors of the earlier naturalists, who were deceived by the external form, and classed the creatures according to shape, and not according to anatomical structure.

To begin, then, with the worms, or annelids, as they are called, being composed of a series of rings bound together by muscular and tendinous substances. The insects and many other creatures, by the way, are also composed of a series of rings; but they possess jointed limbs, and by virtue of those limbs occupy another place in the system of living beings.

The commonest of all the terrestrial annelids is the earth-worm; and there is a marine earth-worm that corresponds with its terrestrial relative in habits and uses. On the sand may often be seen little heaps of contorted sandy strings, possessed of no compactness, but dispersing when touched, and looking as if Michael Scott's familiar were hard at work at his task of twisting ropes from sea-sand, and throwing down his abortive attempts. These ropes or strings are the sand-casts of the lug-worm, a creature that is possessed of no particular beauty, but is very useful to the fishermen, who use it as bait, much as the earth-worm is used by fresh-water anglers. Parties of boys may be seen, armed with spades and boxes, trudging knee-deep in the muddy sand-flats, as soon as the tide goes out, in full search after logs, as they call the worms. Although the shape of this worm is not very beautiful, yet it is not utterly devoid of some beautiful features; for the double row of scarlet branchiæ, or lung tufts that fringe the central portions of the creature, are remarkable for their brilliant tints.

While speaking of this worm, and its representative, the earth-worm, I may as well mention that the popular idea of the multiplication of the earth-worm by division is quite erroneous. The general notion on this subject is that if an earth-worm be cut in two near the middle, the divided portions reproduce those organs which they have lost, and so in a short time the earth is richer by one more worm than before. This notion, however, is untrue. The severed worm seldom seems to recover in the least from its wound, although the portion on which is placed the head survives longer than that to which the tail is attached. The ring next to the wound very soon dies, contracts, withers, and drops off by mortification. The next ring is then attacked in the same way, and dies in its turn. And so on in both portions, the anterior perishing from the wound towards the head, and the posterior portion from the wound towards the tail. The latter portion, indeed, loses the power of locomotion altogether, and can only twist and wriggle about on the spot where it is placed. If only a portion of the tail end be cut off, the remaining part has sometimes sufficient strength to heal the wound, and the creature survives; but the wounded portion is not capable of producing a fresh tail, or even of forming a single fresh ring.

If the sand or stones be carefully examined at low water, certain curious objects will often be found between tide-marks, sometimes existing singly, but generally living in societies. One of these objects is represented on plate F, fig. 2. It is a tube composed of innumerable fragments of shell, or sometimes of entire shells, if they are sufficiently minute, grains of sand, and other similar substances, agglutinated together by a secretion that is poured from the surface of the body, and which soon hardens into a tough membranous substance. The mouth of the tube is adorned with a fringe, which is composed of a number of much smaller tubes formed from the same substances, and in the same manner, as the principal tube. The creature that inhabits this dwelling goes by the name of Terebella. Its empty tubes are sometimes torn away from their attachments by the power of the waves, and in this case are thrown upon the sea-shore together with the algæ, shells, and other *débris* that mark the line beyond which the proud waves can no further go. Generally, however, they are fixed with such firmness, that to procure an entire specimen is a matter of some difficulty. There is no connexion between the tube and its inhabitant, who seems on occasion to be able to take little journeys among the rocks, and even to swim on the surface of the water by spreading its numerous tentacles abroad as the molluscs spread their foot. Sometimes the terebella becomes ambitious, and instead of contenting himself with sand and tiny stones, affixes a stone of some size to his tube. One that I possessed for some time had fastened the centre of its tube to a pebble more than an inch in length, and very nearly the same in width.

The reader will not fail to remark the analogy between the tube-inhabiting annelid, and the larvæ of the common caddis, or stone-fly of anglers, which build tubes in a very similar manner, pressing into the service all sorts of substances, and which, like the terebella and others, has no organic connexion with the tubes in which their soft bodies are sheltered from danger. The tube that is represented in the engraving is of the natural size, and was drawn, as indeed were most of the figures, from an actual specimen.

On the same plate as that which is occupied by the terebella, and at fig 5, may be seen a group composed of innumerable tubes, massed together as if they had been a handful of earthworms compressed into a small space, and then suddenly liberated. These are the tubes of another annelid, called the Sabella, and, like those of the terebella, are built up from the particles of sand on and among which the worm lives. At the bottom of the mass may be seen one of the worms crawling from its tube. These tube-masses may be found in abundance at low water-mark, especially where the corallines are plenty; and the size of the masses is very various, some being only composed of a few tubes twisted together, while others are several feet in diameter. It is not often that a fragment is found where the tubes are so plainly shown as in the specimen depicted. Generally the surface of the mass somewhat resembles a sponge with circular apertures, with here and there a tube, or a portion of a tube, twining itself into the substance. Various algæ are often found affixed to the tubes of these creatures.

Another of these tube-inhabiting worms, or Tubicolous Annelids, to use the correct term, forms a shelly tube so closely resembling that of the ship-worm, *Teredo navalis*, that the two are often confounded with each other, especially if a portion only is in question. This is the Serpula, a group of which is given on plate F, fig. 1, the species being *Serpula contortuplicata*. There are several species of this curious and beautiful worm, one of which, and perhaps the most common, possesses a bayonet-shaped shell, which twists about on the surface of stones or other convenient substances, and does not erect itself freely. But the species that will be more particularly noticed here, after taking a turn or two upon its support, as if to obtain a firmer basis, and at the same time to determine its direction, shoots boldly upwards.

Now if a group of these tubes, situated, we will say, on an oyster-shell, be taken into the hand, they will all appear to be empty and useless; but if the tube is not very much contorted, a something scarlet may be seen at some little distance down the tube, and by that sign the living state of the inhabitant may be known. When the serpula is placed in clean sea-water, it generally remains quiet for a few hours, as if to make itself acquainted with the atmosphere of its new home; by very slow degrees the scarlet object is pushed nearer and nearer to the mouth of the tube, and at last emerges, when it is seen to be an exquisitely formed, conically shaped cork or stopper, its small end being prolonged into a kind of footstalk. Two of these stoppers may be seen on reference to the engraving, but they are hardly represented of sufficient size in proportion to the diameter of the tubes. After a little time, a row of scarlet feathery objects slowly follow the stopper, and in a few minutes spread themselves out into a most elegantly shaped plume.

Slowly as the serpula protrudes itself from the tube, it is by no means slow in retreating. If one of these creatures is fully extended in an aquarium, and the hand is rapidly moved outside without even touching the glass, the worm pops back into its tube with marvellous rapidity, so rapidly, indeed, that the eye fails to follow the movement, and the creature vanishes as if by magic. A cloud passing over the sun, or even the shadow of a person passing by, will have the same effect. It seems evident, therefore, that the serpula must be able to see, although, as yet, no eyes seem to have been discovered. After awhile, however, the creature appears to become partially tame, so to speak, and is less alarmed at a casual movement or shadow. Such at all events was the case with my own specimens, which at first were painfully shy, and avoided all close inspection, but, after a fortnight or so, permitted me to place a lens sufficiently near them to examine the beautiful plumes and stopper.

This last-mentioned organ is the developed one out of a pair which the creature possesses, the other being very small and not put forward to view. This may remind the reader of an analogous arrangement in the tusks of the Narwhal; in which cetacean there are really two tusks, but one only is fully developed, the other lying concealed in the jaw. The beautiful fan-shaped plume is composed of that part of the breathing apparatus which separates the oxygen from the water, and is analogous to the gills of fishes, or the lungs of man.

If serpulæ are kept in an aquarium, they should be closely watched, as they, in common with the sabella and others, have a bad habit of dying when they are not suspected, and so tainting the water, to the destruction of animal life. Most of the tubicolous worms come out of their houses before they die, but the serpula often excepts himself from the general rule, retreats into his shell as far as he can go, and there dies. It is very difficult to discover whether the animal is really dead or only sulky—if the latter, he recovers his temper in a day or so, and waves his plumes as usual; but if the former, a white film begins to collect over the mouth of the tube: this must be accepted as a hint for instant removal. In general, if a serpula does not spread its fans boldly and decidedly from the tube, and permits the stopper to droop over the mouth, it should be gently touched with a camel's-hair brush; if it does not smartly shoot back into its tube, that serpula is in a bad state of health, and must be looked after. It is always better to remove the creature at once, than to run the risk of tainting the water with the unpleasant smell that immediately follows upon the death of any marine animal.

On plate F, fig. 4, may be seen a figure of a creature that does not look at all prepossessing, yet this very animal is one of the most gorgeous creatures that can be imagined, the metallic brilliancy of whose colouring would not suffer in comparison with the plumage of the brightest humming-bird. This animal is popularly known by the name of Sea-mouse, its scientific title being *Halithæa*, or *Aphrodite aculeata*. Edging the body may be seen rows of bristles or hairs which, when simply printed in black and white, give no idea at all of the iridescent colouring of their surfaces; while even, if coloured, the resemblance is but faint, because it wants the changing tints which flash along the hairs whenever they are moved.

It is a strange thing, and one that shows the lavish beauty of creation, that an animal endowed with such glorious colours, that can only be exhibited by a full supply of light, should have its habitation in the mud. When kept in an aquarium, they generally appear to avoid the light rather than to seek it, and keep themselves so hidden among the weeds and stones, that it is not always an easy matter to find them. They are rather migratory in their habits, but not erratic, for they seem to go over the same course week after week; so that, having seen them on one day, it is not difficult to predict their locality on the next.

The bristles of the aphrodite are not only worthy of notice on account of their wonderful colouring, but also on account of their shape. Among other offices, they seem to play the part of weapons, like the spines of the porcupine or hedgehog. But as they surpass the hedgehog's quills in external beauty, so do they in form. There are certain islands, called "Friendly," whose amicable inhabitants are famous for the ingenuity of their clubs wherewith to knock out another friend's brains, and of spears wherewith to perforate him. Many of these spears are made with rows of several barbs, one placed immediately above the other, in order to add more destructive power to the weapon. Now, if the Friendly islanders had possessed microscopes, we might with some justice have said that they took their idea of the many-barbed spear from the bristles of the Sea-mouse; for a magnified representation of one of these bristles would give a very fair idea of the Friendly lance.

All these lances can be withdrawn into the body of the sea-mouse at the will of their owner, and it would therefore be a most unpleasant circumstance, if the creature were to wound itself with its own weapons. In order, therefore, to obviate this difficulty, each spear or bristle is furnished with a double sheath, which closes when it is retracted into the body, and opens again when protruded. It is hardly possible to conceive a more wonderful structure in the whole of the animal kingdom, and certainly not possible to conceive one more beautiful, when the changing tints of orange, scarlet, or azure are taken into consideration.

There is a kind of slimy muddiness about the back of a sea-mouse that rather counteracts the beautiful effect of its hairs. This is caused by the muddy soil in which it loves to reside, and which is strained through a dense mass of fine hairs that interlace with each other, and arrest the muddy particles, while they permit the water for respiration to pass between them. The whalebone plates that fringe the mouth of the Greenland whale have a somewhat similar office, only the fringe of the whale catches molluscs, and that of the aphrodite catches mud.

Wherever rocks are found between tide-marks, their surfaces are usually selected as resting-places by some very curious animals, known by the name of Acorn-shells, which will at once be recognised by the sketch on plate N, fig. 3, where is represented a group of these creatures which have affixed themselves to the shell of a limpet. In the original specimen, the entire surface of the limpet was covered with acorn-shells; but one or two were removed, in order to show the nature of the substance on which they rested. Their scientific name is *Balanus balanoides*, and they belong to a class of molluscs that are called Cirrhopoda, on account of the cirrhi, or ciliated arms, which form their chief characteristic.

The first acquaintance that is usually made with these animals is seldom of an agreeable nature, and generally takes place after the following manner:—An inexperienced but earnest observer is picking his way among the rocks and stones at low water, his eyes being more engaged in searching for curiosities than in looking after his own feet. Suddenly, he puts his foot on a sloping rock, rendered slippery as ice by the slimy algæ that cover it, or is inadvertently caught in a rocky pitfall, whose orifice was concealed by the heavy masses of wrack or tangle that are flung over it by the tide, and at the bottom of which is a pool of water just deep enough to wet his feet, and to irrigate his body by spirting up along the sides of the cavity. In either case he catches frantically at the nearest piece of rock, and finds his fingers cut in several places by the sharp edges of the acorn-shells that have there affixed themselves, and present as uncomfortable a hold for the undefended hand as a wall tipped with broken bottles.

While they are left in the open air, there is nothing attractive in the balani, which seem rather to disfigure the rock than to improve its appearance. But when the sea returns and brings back the welcome supply of nourishment, these dull, lifeless objects suddenly start into activity, and begin to fish as industriously as if they knew that they had only a limited time for eating, and must, during that time, procure a sufficiency of food to employ their digestive organs while the tide is out. The manner in which the Cirrhopoda fish is very remarkable. Some animals, like the sea-anemones, hang out a net and await the approach of prey, who unwarily come within the scope of their power, and so rush to their own destruction. Other creatures hang out fishing lines, like the common fresh-water Hydra, or the beautiful marine Beroe, which will be described on a future page. Others, again, chase their prey through the water, and capture it by virtue of superior swiftness or cunning. But neither of these modes is employed by the balanus, which is furnished with a veritable casting-net, which it ever and anon throws expanded into the water, and then retracts when closed. The action of a man throwing an ordinary casting-net is much the same as that of the acorn-shell. If the reader will refer to the plate, he will see two of these creatures in the act of making their cast, and the net is formed from the cirrhi from which the entire class derives its name.

Each cirrhus is found, on examination, to be double, the pair springing from a single footstalk. They are of a partially horny consistence, and separated into numerous joints, each joint being furnished with long stiff hairs. These hairs stand out boldly from the centre, and the consequence is, that when the whole apparatus is fully extended, it forms a kind of network of hairs, permitting none but the smallest substances to pass between them. In the balanus, the cirrhi are of a delicate white colour, and have a singularly elegant appearance as they are alternately thrown abroad and gathered together again. There is hardly a prettier sight than a large stone, or piece of rock, that is covered with balani, and immersed in clear sea-water. Each little conical shell opens at the tip, and from the aperture a fairy-like little hand is constantly thrust, grasping at some coveted object, and then closed and withdrawn. There is a grace and elegance about the whole movement that is not easily described. This sight may often be witnessed in the rock-pools when they are of sufficient depth to cover the balani, and are not exposed to the action of the wind.

With all their beauty, however, the balani are uncongenial inhabitants of an aquarium, although they add much to its appearance at first. They soon become languid, their graceful cirrhi remain half protruded from the shell, they then die, and shortly exude such a detestably scented gas, that the surrounding water soon becomes unfit for the respiration of the other inhabitants—they in their turn die, and the whole aquarium is ruined. So, present beauty must be sacrificed to ulterior service; and if any balani are growing on a rock intended for the aquarium, they must be removed before it is placed in the tank.

They do not seem to be particular as to their place of residence, for they may be found on rocks, wooden piles, stones, and even on living shells, of which they most affect the limpet, because it is not of migratory habits.

It is a very remarkable fact, that although the balanus never moves from the spot on which it has taken up its habitation, and, indeed, is incapable of any kind of locomotion, yet when very young it was an active, wandering little creature, furnished with jointed limbs, much resembling a young shrimp or crab, and swimming freely through the water with a succession of bounds. When first discovered, the young balani were thought to be veritable crustaceans; but after careful observation they were seen to affix themselves to the sides of the vessel in which they were placed, and straightway to change their roving life for an existence of settled quiet. Similar strange developments take place in many marine animals, but there will not be sufficient space for their discussion.

The balanus has a very near relative going by the popular name of Ship-barnacle, and the scientific title

BARNACLE.

of *Pentalasmis anatifera*, the latter title signifying "the five-plated goose-bearer." It is called Pentalasmis, or five-plated, because its shell is composed of five distinct portions, curiously arranged, and between them the cirrhi are protruded. The word "goose-bearing" is given to it because an old writer named Gerard, who lived in 1636, discovered that the Bernicle-goose (*Ber-*

BERNICLE GOOSE.

nicla leucopsis) was produced from the ship-barnacle; and in order to prove his own rather startling account, he gives drawings of the creatures in all their stages, from the mollusc to the bird. Whether the worthy man intended to deceive, or was himself the victim of others, it is impossible to say. His account is so quaint that I here give an extract:—

"What our eyes have seen, and hands have touched, we shall declare. There is a small island in Lancashire called the Pile of Foulders, wherein are found the broken pieces of old and bruised ships, some whereof have been cast thither by shipwracke, and also the trunks and bodies with the branches of old and rotten trees, cast up there likewise; wherein is found a certain spume or froth, that in time breedeth into certaine shels, in shape like those of the muskle, but sharper pointed, and of a whitish colour; one end whereof is fastened into the inside of the shell, even as the fish of oisters and muskles, the other end is made fast into the belly of a rude masse or lumpe, which in time commeth to the shape and form of a bird: when it is perfectly formed the shell gapeth open, and the first thing that appeareth is the aforesaid lace or string; next come the legs of the bird hanging out, and as it groweth greater it openeth the shell by degrees, till at length it is all come forth and hangeth only by the bill: in short space after it commeth to full maturitie, and falleth into the sea, where it gathereth feathers, and groweth to a fowle."

The anatomy of this barnacle is curious, and will repay examination. The shell should be removed, and the animal carefully displayed with the assistance of a pair of scissors and a needle or two: all dissections of small animals should be made under water; or if the dissected creature is intended to be permanently preserved, it should be immersed in the fluid which is used as a preservative.

These creatures are often found clinging in great numbers to the bottoms and keels of vessels, sometimes interfering with their speed. Their growth is very rapid, and it has often happened that a ship has started upon a short voyage without a single barnacle adhering to her planks, and yet has come back encumbered with a whole army of them. They are often found adhering to pieces of wreck, or to floating spars that are cast upon the shore by the waves. The stalks or tubes of the individual represented in the engraving are not sufficiently long in proportion to the dimensions of the shells themselves, and ought to have been drawn nearly double their length.

After a gale, especially if the wind sets landwards, the shores afford a great harvest to the naturalist; and if the gale and the spring-tides coincide, he inwardly wishes for the hundred eyes of Argus to look after the objects that lie scattered on the shore, and for the hundred arms of Briaræus wherewith to pick them up. Among the strange things that are cast on the shore will be seen many lumps of a jelly-like substance, varying much in size, called popularly by the name of Jelly-fish. These are strange creatures, of wondrously low organization, that thickly populate the ocean, and are anything but shapeless when living and in health. Scientifically they are called *Acalephæ*, from a Greek word signifying a nettle, because many of the species have the power of stinging the hand that incautiously touches them.

Sometimes they lie on the shore in vast numbers; and there is a story on one of our coasts, that a farmer ordered down his carts to the sea, and carried away several cartloads of jelly-fish to serve as manure for his fields; but by the next morning the heaps of jelly-fish had disappeared, leaving behind them a few lumps of membranous threads. In fact, all the real animal matter that the carts had carried to the fields might have been conveyed in the farmer's own hand, for jelly-fish are really little but animated sea-water.

On plate N, fig. 2, will be seen a singularly shaped creature, bearing two long threads covered with spiral tendrils; this is one of the jelly-fishes, called by the name of Beroe or Cydippe, and a wonderful creature it is. If on a calm day a gauze net is passed gently through the water, there will often be found adhering to its sides sundry little gelatinous knobs, perfectly transparent, and apparently lifeless. Now, if the net be lowered into a glass vessel of pure sea-water, and slightly agitated, the lump of jelly will be loosened, and left in the water. For a moment the eye fails to perceive that the water has any inhabitant at all; for the beroe, as the gelatinous knob turns out to be, is itself little but sea-water, but may soon be recognised by the flashes of light that appear on its surface. It is a creature that can hardly be drawn, for it ought to have no outline, and only to be shown by the brilliancy of its surface, which surpasses that of the water around. Presently, as the creature begins to feel more at home in its new habitation, it swims about with an easy gliding movement, and an iridescent light shows itself on one part of the surface. The iridescence continues to increase, and at last is seen to reside in eight longitudinal bands that completely encircle the animal; over these bands the light plays, and at last all the colours of the rainbow ripple over its surface with indescribable beauty.

These iridescent bands are the organs of locomotion, and it is to their form and mode of use that the beautiful colour is owing. By the side of the beroe may be seen a magnified portion of one of these bands. Its surface is covered with little square scales, disposed in a manner somewhat similar to the boards of a water-wheel. Each of these steps, so to speak, is capable of motion backward and forwards, and by their rapid and successive motion a series of prisms are formed, and by them the light is decomposed to the prismatic colours; this iridescence is best seen when the sun shines upon the creature.

When the beroe has been watched for a little time as

it swims about in its glass prison, two long and most delicate threads will be seen depending from its exterior, and falling into graceful curves as the creature ascends or descends in the water. The threads are so exceedingly delicate that they are not observed at a first glance; and when they are seen, rather convey to the spectator's mind the idea of spun glass, than of any animated structure. Indeed, the whole creature looks as if it were formed of crystal, cut and polished, and the threads almost seem to be spun from its substance as it moves about. These threads are called the fishing-lines, and if closely watched are found to be fringed with smaller tendril-like threads, that are dispersed along the chief line, just as a fisherman attaches several baits to his line by supplementary strings. The fishing-lines can be entirely withdrawn into the body of the animal, or they can be shot out to lengths that appear wonderful, considering the size of the creature to which they belong. The supplementary tendrils elongate themselves when the fishing-line is drawn out to its full length, and become more tightly twisted as the line is retracted into the body. Many people, and especially those who live on the sea-shore, imagine that the beroe is the egg of the sea-urchin.

It must be remembered that the creature can alter its shape by expansion and contraction; which circumstance accounts for the fact, that if several artists sketch this creature, each figure may have a different form. The species represented in the engraving is *Cydippe pileus*, shown as it appears when fully expanded. The life of the creature is fragile as its form; and if it is kept in a vessel of water, it plays about rapidly for a time, then dies, and disappears as if it had melted into nothing. Yet, if it be cut into pieces while lively, or broken up by the force of the waves, as is often the case, its ciliated bands still continue to perform their work, and the iridescent light plays over the fragments as beautifully as when the creature was entire. It is seldom or never found in an entire state at the surface of the water when the wind is rough, but sinks below into the calmer regions, where its delicate organization is not exposed to the rude collision of wave and wind.

The beroe glides along by means of the ciliated bands; but this is only one of the means of progression employed by the Acalephs. Some move themselves about with cirrhi, and are therefore called Cirrhigrade; the beroe being called a ciliograde. Others again are named Physograde, because they are buoyed up by a kind of bubble, or bladder filled with air. The well-known Portuguese Man-of-war is a good example of this order. There is another order which moves through the water by a series of regular pulsations like those of the lungs, and the species belonging to it are called, in consequence, Pulmonigrade. An example of a pulmonigrade Acaleph is given in plate N, fig. 1. It belongs to the genus Ægerea, of which many species may be found on our coasts, and, with many others, passes under the general title of Medusa. The size of these creatures varies excessively, and there are strange peculiarities in their growth and structure, which should be learned from some one of the books devoted exclusively to the Acalephs. The movements of a little Medusa in a clear glass vessel are exceedingly graceful, and may easily enough be witnessed; as if a vessel is filled with water drawn from the surface of the sea on a calm day, there are generally a few Medusæ in it. But if there should be none, a little work with a gauze net will secure plenty.

Very many of these Acalephs are phosphorescent, and through their instrumentality the sea appears at night as if filled with fire. This property is especially shown when a little breeze ripples the surface, or a boat dashes the water aside. In the latter case the oars appear to throw from them torrents of fire, and a blazing line marks the direction which the boat has taken. The chief cause of this phosphorescence is a very tiny creature, called *Noctiluca miliaris*. If a vase be filled with sea-water and placed in the dark, it will emit small sparkles of light whenever it is tapped, or even when the foot is stamped on the ground. This phenomenon is more exemplified at the sides of the jar, and but a very few sparks are perceptible at the bottom. Each spark is caused by a Noctiluca; and if it is wanted for examination, it may be caught in a glass tube, in the manner employed for microscopic animalcules, and brought under the necessary magnifying power. Its natural size is about half that of a common mustard-seed. When submitted to magnifying power, it appears to be a creature of a form nearly round, but with a notch or depression on one part of its circumference. Close by that depression is a little knot of muddy matter, from the centre of which springs a kind of tail, or perhaps proboscis, by the agitation of which the creature rows itself about in the water. The entire form of a Noctiluca is not unlike that of a melon, the proboscis being the stalk.

CHAPTER IX.

CRABS—LOBSTERS—SHRIMPS—PRAWNS, AND FISH.

Among the living creatures that force themselves on the notice of any one who walks on the borders of the sea, the various crustaceans are perhaps the most conspicuous. Without attempting here to treat of the Crustacea scientifically, I shall mention those creatures that may be seen almost on any day and almost on any shore, leaving the deeper scientific details to be obtained from the very elaborate works that exist on the subject.

It is nearly impossible to walk for more than a few paces on the wet space between tide-marks, without disturbing a host of little crabs, that scuttle about in dire perplexity, either trying to flatten themselves against the ground, hoping to be mistaken for pebbles, or endeavouring to conceal themselves under the shade of a bunch of wrack. Sometimes, on lifting up a heavy mass of sea-weed, out comes a crab very unexpectedly, holding up a pair of claws with so ferocious an air that he often escapes before his discoverer has quite recovered his presence of mind. The former species, that try to escape in such a hurry, are generally the uneatable green crab, although even it is often found of no small size; while the big pugnacious one is of the edible kind, and, from his objection to capture, seems to know it. Sometimes crabs of a very tolerable size may be found concealed in the crannies of the rocks, where they are concealed by the rock itself, and by the fuci and laminaria that hang about in great profusion. Barelegged boys may be often seen creeping about among the rocks, and armed with a basket and an iron rod hooked at the end. This latter weapon is ever and anon thrust into the holes and clefts of the rock; and should an unfortunate crab have there concealed himself, he is soon hooked out of his retirement, and, if edible, consigned to the basket.

The great monsters that are brought to market are mostly caught in sunken baskets, much on the principle

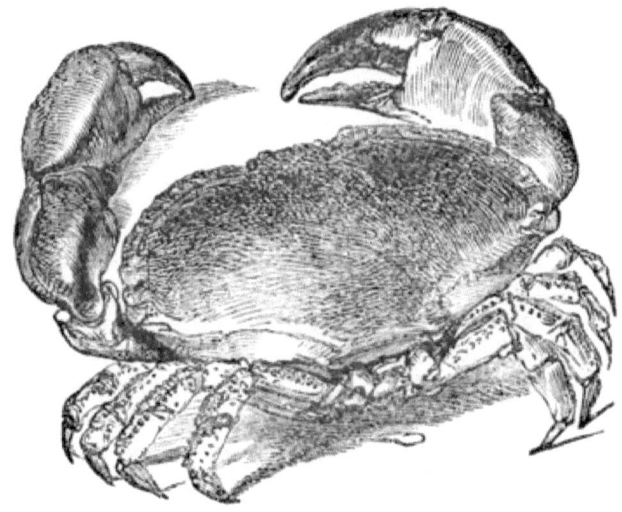

CRAB.

of the mousetrap, which permits an animal to enter without any difficulty, but opposes an effectual barrier to his egress. It is not always safe to grope for crabs, unless a companion be at hand, for a powerful crab has actually been known to grasp with its claws the hand of its opponent, and to hold him there without the power of moving until a passer-by came to his assistance. Should such a circumstance occur, the best plan for making the animal loosen its hold is said to be by detaching the claw that is unemployed.

The young of the crab is quite unlike the adult animal, and has been described under the name of Zoea.

In this state it is a very quaint-looking creature, is possessed of a long tail, two great eyes, something like those of a diver's helmet, and wears a spike on its throat nearly as long as its entire body. It is no marvel that it has been treated of as a separate creature from the crab, for it bears about the same resemblance to the crab that a caterpillar bears to a butterfly, or a wire-worm to a beetle. The long tail of the Zoea formed one of the distinctive points that separated it from the crabs; and yet, if a crab is laid on its back, a tail is seen to be tucked up under its body, in a position something like that assumed by the tail of an alarmed dog.

Many species of crabs may be found on the sea-shore; that is, if they are sought in the proper localities. The two species already mentioned are totally incapable of swimming. They can crawl upon the shore, half bury themselves in the sand, or push their way among the algæ with much rapidity; but if they are thrown into deep water, they sink helplessly to the bottom, spreading about their limbs in the vain search after some object which they can grasp. There are, however, several species of crabs found on the British shores, which are good swimmers, one of which is given on plate M, fig. 4. This is the Velvet Swimming Crab, or the Velvet Fiddler, as it is sometimes called. If the figure of this animal be compared with that of the common crab, on p. 112, the reader will observe that there is a considerable difference between the two creatures; one of the chief distinctions lying in the shape of the hinder pair of legs, which in the common crab are sharp and rounded, but in the swimming crab are flattened at their extremities. These flattened limbs are used as oars or paddles, and by their repeated strokes the creature is able to urge itself through the water with some velocity. The peculiar movement of the limbs being thought to resemble the action of a violinist's arm, the crab has thence derived its name of Fiddler.

The Velvet Fiddler (*Portunus puber*) is common

I

enough on our coasts, and is a tolerably hardy inhabitant of an aquarium, but not a safe tenant. It is a most ferocious creature, lurking unseen in a corner, and from its den darting forth at any unsuspicious inhabitant that may come near. One of these creatures has been known to attack a moderately-sized hermit crab, and to destroy it with a single snap, despite its shelly habitation and strong claws. After having killed the poor hermit, the fiddler proceeded to eat its body.

There are some crabs, again, which, in consequence of their peculiarly shaped body and long sprawling limbs, are termed Spider Crabs; for an example of which creatures, see plate M, fig. 3. The bodies of these crustaceans are short, wide, and produced into a snout-like form in front. If they will live, they are useful creatures in an aquarium; for they are good scavengers themselves, and, in addition, often carry on their shells a whole army of zoophytes. Two of these creatures, which inhabited a large aquarium belonging to a friend, were perfect treasures to the microscopist; for when a specimen of a living zoophyte was wanted, one of the spider crabs was hunted up, and the requisite portion removed. They were both rather sluggish animals, and generally resided in two dark holes, in different parts of the aquarium, where a practised eye was needed to discover them.

The species represented is *Maia squinado;* but there are several other spider crabs to be found on the shores, some of whom possess limbs so wonderfully elongated, that they seem to have been subjected to the process of wire-drawing. These may be captured at spring-tides, when the water has sunk much below its usual level, and left the unsuspecting crabs on dry land. Various curious fish, and other creatures, may also be taken at the same time.

We now come to a very curious race of creatures, the Soft-tailed Crabs. Those already mentioned are entirely covered with a strong shelly mail; but there are others whose tails are left bare and defenceless, and which are forced to seek an artificial defence in lieu of natural armour. These creatures are generally called Hermit Crabs, because each one lives a solitary life in his own habitation, like Diogenes in his tub; and sometimes go by the name of Soldier Crabs, on account of their very pugnacious habits. The species here given is the Common Hermit Crab (*Pagurus Bernhardus*),

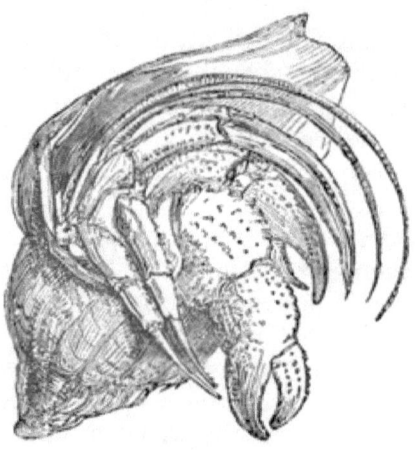

HERMIT CRAB.

and the particular individual is inhabiting a whelk-shell, a domicile that is in great request when the creature grows to any size. Hermit crabs may be found plentifully on the coasts, of all sizes, and inhabiting all kinds of shells, from the trochus to the whelk; and I have often seen a poor little hermit forced to take up with a huge whelk-shell, of which only the last whorl remained entire, and into which he exactly fitted. He was almost at the mercy of his habitation, for he could not hold it against the power of the waves, which tumbled it over and over most ruthlessly, while the hermit was making futile grasps at stones and sand by which to arrest his progress.

The hermit is furnished with an apparatus of pincers at the extremity of his tail, by which he holds firmly to the shell in which he takes up his temporary habitation, and he flattens himself so firmly against the shell that it is difficult to seize the creature at all; and even when a grasp of any portion can be secured, the hold of the tail is so firm that the animal runs some risk of being torn apart, sooner than leave the shell. Some years since, I was rather anxious to see how the hermit got into his shell, and so, having caught a tolerably large one in a whelk-shell, I tried to pull him out. However, he stuck so close to his shell, that there was no hope of success without inflicting much injury, and I should probably have let him escape, had not an idea then come across my mind. Close by the rock-pool where the hermit had been captured was a colony of very fine sea-anemones (*B. Crassicornis*), and I thought that, probably, by their aid, Mr. Hermit might be enticed out of his shell, even if he would not be dragged out. So I popped the hermit among the wide-expanded tentacles of the crass, which immediately began to contract. The hermit was evidently acquainted with his danger, and, in his hurry to escape from the adhesive tentacles that were twining about him, loosened his hold of the shell, and was instantly plucked out. I let him walk about for a while in the pool, where he looked very woe-begone, trailing his defenceless tail behind him as if he were ashamed of it. After a while I dropped a damaged Purpura shell into the pool, and the crab at once went up to it, and after a very short examination, stuck the end of his tail into it, for it was not large enough to accommodate the entire tail, and walked about as before. At last, I put the original habitation into the pool, to the very great delight of the hermit, who exchanged shells with a marvellous rapidity, and seemed so much at home again that I could not think of disturbing him. I have frequently tried the same plan of enticing the hermits out of their shells, and never found it to fail.

The combative propensities of these creatures are quite wonderful for their development. If only two hermits of tolerably equal size are placed in an aquarium, they are not content with appropriating different portions of the vessel to themselves, but must needs travel over it and fight whenever they meet. This struggle is constantly renewed, until one of them discovers his inferiority and makes way whenever the victor comes near. When they fight, they do so in earnest, tumbling over each other, and flinging about their legs and claws with singular energy.

They are not at all particular about diet, so that it is of an animal substance, and will eat molluscs, raw meat, or even their own species. More than once, when a hermit has died, I have dropped the body into the water so as to bring it within view of another hermit. The little cannibal caught the descending body in one of his claws very dexterously, and holding it firmly with one claw, he picked it to pieces with the other, and put each morsel into his mouth in a rapid and systematic manner, that was highly amusing. It was literally "tucking in." Only the soft abdomen was eaten, and the hard legs, claws, and thorax rejected. Some prawns came and tried to eat the rejected portions, but unsuccessfully, although they dragged them about the water for a few minutes.

When one of these hermits is in captivity, and feels ill, he crawls out of his shell, and generally dies in an hour or two afterwards. It is a very curious propensity, and one that is shared by the tube-inhabiting worms, as has already been mentioned. The habit is the more remarkable, as the usual instinct of animals leads them to the most retired spots that they can find, and there to resign themselves to death.

The formation of the hermit is wonderfully suited to the strange habitation which it adopts, and a hermit when removed from the shell would hardly be recognised for the same creature as that which was snugly curled up within it. Even the claws are modified, for the purpose of lying smoothly in the shell's mouth, one of them being very large, and placed in front, as a shield and weapon united, while the other is very small, and is almost wholly retracted within the shell.

When a hermit desires to change his habitation, he goes through a curious series of performances, which, if he had hands, we should be disposed to call manipulations. A shell lies on the ground, and the hermit seizes it with his claws and feet, twists it about with wonderful dexterity, as if testing its weight; and having examined every portion of its exterior, he proceeds to satisfy himself about the interior. For this purpose, he pushes his fore-legs as far into the shell as they will reach, and probes with their assistance every spot that can be touched. If this examination satisfies him, he whisks himself into the shell with such rapidity, that he seems to have been acted upon by a spring. Such a scene as this will not be witnessed in the sea, unless the hermit is forcibly deprived of his shell, as in the cases above mentioned; but when hermits are placed in a tank or vase, they seem to be rather given to "flitting."

The crab is not always the sole inhabitant of the protecting shell, for one of the curious worms, called Nereides, is very frequently found in joint possession. The fishermen value the Nereis for bait, and turn out the poor hermits in order to obtain this worm that is concealed within the whelk-shell.

The Common Lobster (*Astacus marinus*) is an example of another order of these creatures, the Macroura, or Long-tailed Crustaceans. The cray-fish, prawns, and shrimps, all belong to this order. The lobsters are usually taken in basket-traps, after the same manner as the crabs, and when caught they are kept in well-boxes, through which the sea-water runs freely; that is, if there is no immediate sale for them.

REPRODUCTION OF LIMBS.

Both crabs and lobsters, together with other crustaceans, possess the remarkable faculty of throwing off a limb or two when injured, or even if they are alarmed,

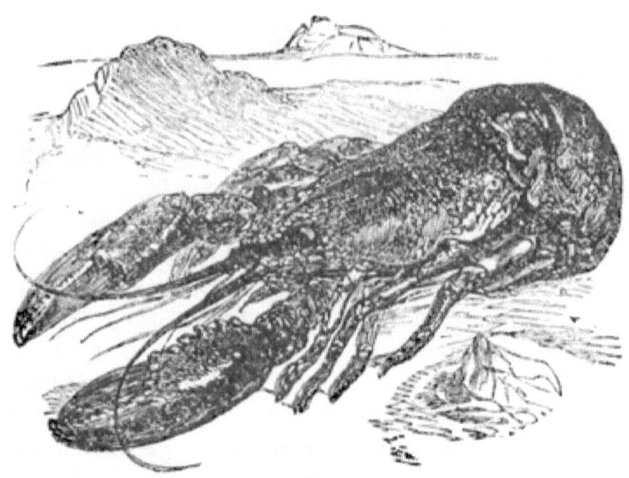

THE LOBSTER.

and of reproducing the lost members. This voluntary amputation often takes place even in the huge claws that lobsters carry, and which are so valued at table. Sailors sometimes hold out a threat to the fishermen, that they will sail to the lobster-grounds, and there discharge heavy guns; the effect of which would be to make the lobsters throw off their claws, and thereby render them unsaleable.

Most crustaceans are pugnacious in character, and it often happens that when they fight, they inflict serious injury upon each other's limbs; and in such cases the maimed limb is detached, not at the wounded spot, but at the joint immediately above, and after awhile a slight protrusion opens itself at the amputated joint, which protrusion becomes more protruded, and in short space develops into a limb. It is a very common circumstance to find a lobster with one very large claw, while the other is of comparatively small size. There is a very wise object in this power. The blood vessels of the crustaceans are but slightly contractile, and in consequence, if a wound is inflicted, the vessels continue to bleed freely. But by this amputation the wounded surface is reduced to a very small portion, and the substance of the joint contracts with sufficient force to stop the bleeding.

The shell of these creatures is formed of unyielding calcareous substance; and although it may be a most excellent defence for the full-grown crab or lobster, it leaves no room for growth. In order to obviate this difficulty, the crustaceans are possessed of the power of discarding their shells at certain seasons of the year; at which time, also, a new and larger shell is formed. This, in its turn, is cast off; and so continually, until the creature has attained its full growth. Not only are the mere shelly coats of the body and limbs thus changed, but the lobster, for example, when it changes its shell, discards with the shell the following portions of the body:—

The footstalks of the eyes.
The external cornea of the eyes.
The internal thoracic bones.
The membrane of the ear.
The membranous covering of the lungs.
The tendons of all the claws.
The lining of the stomach.
And the stomachic teeth.

The lobster seems to experience some difficulty in casting its skin; and from this list of organs that are changed, it is but little wonder that there should be a difficulty.

There are certain glass toys that were once much in vogue. A reel covered with thread, or a model of a railway engine and tender, were exhibited in a bottle with a very narrow neck. The puzzle was, to discover how the reel got into the bottle. But in the case of the lobster, there is a puzzle of precisely an opposite character: for it is easy enough to understand how a lobster gets into his shell, or rather how the shell forms over the lobster; but how he gets out of it, and especially how he withdraws his huge pincers without leaving the slightest mark of fracture, is a riddle far more wonderful than that of the bottle. At all events, it is accomplished, and that with such nicety that the cast shell exhibits precisely the same appearance as when it surrounded the living animal.

When the shell is rejected, the creature is almost undefended, its body being only covered with a membranous skin; and is therefore liable to be injured by foes who would be treated with contempt were the shelly armour in its place. So the defenceless animal conceals itself in a quiet spot, and there waits until another shell has been secreted.

The movements of the lobster, and indeed of all its relatives, may be reduced to two kinds, crawling and shooting. Its legs are the means by which it accomplishes the former mode of progression, and the tail by which the latter. I should rather have said, that the tail was the organ of retrogression; for when the lobster shoots through the water by means of the tail, it is in a backward direction. The extremity of the tail is furnished with an array of broad plates, so disposed, that when the lobster violently bends its body into the curved shape that it assumes when boiled the force of its action against the water is so strong, that a single stroke will urge the creature to a distance of twenty feet or more, and even enable it to spring out of the water. The natural position of the lobster is straight, and it only curls itself on emergencies, such as a sudden fright, or immersion in hot water. And its sight is so good, or its instinct so wonderful, that it can thus throw itself between two rocks where is barely room for its body to pass. As to the sight, it may well be good, for the lobster possesses compound eyes like the insects, only the shape of the lenses is square instead of hexagonal. Many crustaceans have their eyes hexagonal instead of square.

Beside the crab, the lobster, and the crayfish, two other crustaceans find their way to the table; some of them gaining in liveliness of hue by their passage through the hot water, such as the lobster, the shrimp, and the crayfish; while some, as the crab, make very little change, one way or the other, and some, as the prawns, positively lose their exquisite tints. Heat applied in any way has the same effect, and so has alcohol.

As the shrimps are of more common occurrence than the prawns, they will have the precedence. Still, many so-called shrimps are really prawns, and among them may be noticed the red shrimp, as distinguished from the brown. The true shrimps are called by the fishermen "sand-raisers," and with reason; for they have a habit of scooping out furrows in the sand, and sinking into them until hardly any portion of their bodies is visible. And in doing so, they raise quite a cloud of sand, which settles down again and assists in obscuring their bodies.

They are caught in peculiarly-shaped nets, which are pushed along the bottom of the sea, and into which the alarmed shrimps rush. The shrimp net, however, contains many objects besides shrimps or prawns, and it is useful to bargain with a shrimper to put all his "rubbish" into a basket and bring it back at the end of his short but hard work. In this way all sorts of marine creatures are captured, especially the smaller fish, and various crustaceans. It is also a charity to the shrimpers, for their work is severe, very trying to the constitution, and badly paid. They are mostly good-natured, honest fellows, always ready to earn a shilling, and can be made into useful assistants as soon as they have got over their astonishment at the value attached to the objects which they have been in the habit of throwing away as rubbish, and which they will still throw away if not looked after.

In the little rock-pools that are left by the tide, numbers of small shrimps may be found, but not easily caught; for they are exceedingly active, darting about with the rapidity of arrows, and, being just the colour of the sand, are difficult to perceive. They cause innumerable furrows in the sand by their little darting flights; and although their marks are usually obliterated by the returning tide, yet there are instances where the furrows, together with the spots produced by rain-drops, and the footmarks of shore birds, have been petrified, and remained as witnesses of events that took place many ages ago. If these little shrimps are desired as specimens, they may be easily captured, by passing a gauze net rapidly through the water. The shrimps are startled at the flash of the net, and as they dart about wildly, here and there, are caught in the very object which they were endeavouring to avoid.

The shrimp is a prolific creature, and produces a large number of eggs, which it carries about until hatching time comes. The young shrimps are comical little creatures, anything but harmonious in their proportions, and bear no more resemblance to their parents than a newly-hatched blind chick does to a gallant game cock. They seem to be gregarious in their habits, and at a little distance look like a cloud of active white particles. They can be easily brought together for observation, or captured, by holding a lighted candle to the side of the vessel which contains them; for they crowd to the light like so many moths or gnats, which latter insects they slightly resemble. Unfortunately, many marine animals are very fond of young shrimps, and a great amount of catching and eating goes on whenever a fresh batch of shrimps comes into existence, so that only a very small per centage attain maturity.

The prawns when living are most exquisite beings, their partially transparent bodies being diversified with delicate tintings, and their radiant eyes glowing like living opals. A boiled prawn loses as much of its living beauty as a trout or a mackerel when subjected to the same process. I have already remarked that the compound eyes of the crustacea are analogous to the corresponding organs in the insects, and the analogy is further carried out by the power of reflecting and refracting light. The eyes of many insects, especially those who fly by night, possess this property; and of all British insects, the Common Death's-head Moth is perhaps the most conspicuous in this respect. Either by daylight, or when the rays from a candle fall upon the eyes, they glow as if lighted from within by some hidden fire; a circumstance which adds in no small degree to the terror which is often inspired in the uneducated mind at the sight of one of these insects. In all cases the light departs together with the life of the animal: its origin is not as yet clearly ascertained. Even the eyes of the common Dragon-fly, a diurnal insect, possess a kind of fiery glow when viewed during the life of the creature, but turn to dull, dead hemispheres as soon as it perishes. The light that is reflected from the eyes of cats, &c. is accounted for on principles that do not hold good with regard to the compound eyes of insects or crustacea. The ordinary edible prawns are not found between tide-marks, except occasionally when an unhappy individual has been driven towards the shore, and has not been able to regain the sea before the waves have retired. For this creature, the shrimping net or the dredge is requisite; and as it is tender in constitution, a vessel of sea-water should be ready for its reception when taken out of its native haunts.

There are, however, several species of shore prawns which quite equal, if not excel, those of the deeper waters in beauty. One of these, the Common Æsop Prawn, is given on plate M, fig. 1. It is called the Æsop Prawn because it wears a kind of hunch upon its back, thereby following the example of the great fabulist. Its scientific name is *Pandalus annulicornis,* or the Ring-horned Pandalus. The title "ring-horned" is given to it, because the horns, or antennæ, are exquisitely ringed with scarlet lines at regular distances. These antennæ are most lovely organs; and as the prawn swims through the water in its usual graceful gliding progression, the antennæ wave to and fro, producing elegant and ever-changing curves. The whole body of the creature is covered with scarlet lines, which show out exquisitely upon the pellucid groundwork.

These creatures will not be found in the winter, or even in the early spring; but in the summer months they may be seen in abundance in the rock-pools, and captured by means of the gauze net without any difficulty. If the pool is too large, and permits the enclosed animals to escape from the net by means of their extreme activity, the water may generally be drained away by a judiciously cut channel, well guarded by stones and pebbles, or even by the more simple but more tedious mode of baling; the collecting jar makes a very good baling pan. By adopting either of these plans, the surface of water soon becomes contracted, and the imprisoned animals are driven into narrower limits, from which they may be extracted at leisure.

On most sandy shores a curious appearance may be seen bordering the skirts of the waves, an appearance as if innumerable little grasshoppers continually leaped into the air, and in some places so numerous as to fill the air with a sort of misty cloud, to the height of several inches from the ground. Often as the promenader walks along the sea-shore, his footsteps put up whole swarms of these creatures, and induce him to catch them, or rather to attempt their capture. Perhaps one very large individual jumps into the air, and comes down so determinately, that it is marked out for a victim. Down comes the hand upon the spot, but the creature has actively hopped away, and is making off with a succession of agile leaps, that remind one of a kangaroo or a bull-frog. If the pursuer can drive the agile creature from the sea, he may run it down after a smart chase; and when he has caught it he will see that it is a little crustacean, whose form may be recognised on plate M, fig. 2. From its hopping propensities, it goes by the name of Sand-hopper, or Sand-skipper.

It generally lives on the shore, burrowing deep holes in the sand, where it lies concealed until the waves again cover the sands. And if fine specimens are wanted for collection or preservation, they may easily be obtained by digging into the sand with those wooden spades, of which there is no lack wherever there are children, and so pouncing on the sand-skippers before they can recover their alarm at so sudden an entrance into the light of day. They may also be found plenfully swimming about in the rock-pools, or concealed in the masses of ulva or enteromorpha that mostly fringe those miniature ponds. If a basketful of these weeds be plucked at random, and then thrown into a large vessel of sea-water, some twenty or thirty sand-skippers will generally be seen swimming about, and may so be captured.

They feed on the green sea-weeds, and would be hurtful inhabitants of the aquarium did they not serve as food for the anemones, crabs, and other living creatures that are generally kept in such receptacles. It is surprising how soon they vanish from the scene, as soon, indeed, as a stock of carp and roach vanish if placed in a pond where several large pike have taken up their abode. A whole handful of sand-skippers may be transferred to a well-stocked aquarium, and in a week or so hardly one will have survived; there will be plenty of empty shells and rejected limbs at the bottom of the aquarium, but nothing more than their vestiges to tell that sand-skippers once were.

In the same rock-pools where the shrimps, prawns, and sand-skippers are found, there reside also temporarily numbers of little bright eyed active fish, hardly distinguishable from the shrimps until captured. One of these fishes is shown on plate N, fig. 5, and its name is popularly, the One-spotted Goby, and scientifically, *Gobius unipunctatus.* It derives its name from the single spot that may be seen on the dorsal fin, and which is so conspicuous a mark that by it the creature may be easily distinguished, at all events with sufficient accuracy for ordinary purposes. There is another species of goby, called the Two-spot, that is very common on the coast; so common indeed are these little fish, that I have taken upwards of thirty in as many seconds, merely by sweeping with the gauze net the waters of a rock-pool that had been condensed, as it were, by draining.

The gobies are hardy little fishes, and are able to withstand the prejudicial influences that are inseparable from even the best regulated aquarium. Some three or four are sufficient in number, and impart to the tank a liveliness that is very pleasing. The ventral fins of the gobies are so formed that they can be pressed together and used as a sucker, by means of which they can adhere firmly to the glass forming the sides of the aquarium, or to the rocks and stones of their native sea. The rapidity, too, with which a goby affixes itself to the glass is quite surprising. These little fish are terrible enemies to the shrimps, for they feed greedily either on the eggs themselves, or on the young shrimps when they have just emerged from the egg. They also feed much on the animalcules of various kinds that throng the alga, and so may be conveniently fed by placing in the tank a handful of freshly gathered ulva, enteromorpha, or indeed any of the sea-weeds whose growth is sufficiently dense to afford shelter to the animalcules. By the aid of a lens, the tiny creatures may be seen coming by thousands out of the floating sea-weed, and snapped up almost as fast as they show themselves.

The Black Goby (*Gobius niger*) may also be captured

as he lies lurking in cunning recesses beneath the stones and rocks, waiting for prey. He is decidedly a fierce fish, and its admission into an aquarium is a doubtful point, inasmuch as he has been known to catch and devour the two-spot goby. It is a larger fish than either of those already mentioned, being about three inches in length.

Another very curious fish is found in much the same locality as the gobies. This is the Shanny, Tansy, or Smooth Blenny, as it is indifferently named, one individual of a large family, whose features are suficiently remarkable for recognition. The scientific name of this fish is *Blennius pholis*, and a portrait of it is given on plate N, fig. 6.

Any one who possesses an aquarium should search after this creature, for it is quite as hardy, if not more so, than the gobies themselves, and is also a bold active fish, making itself very comfortable in its new home, and sparing no opportunity of procuring food, even snatching it from the very jaws of less active fish. The colour of this fish is variable, some specimens being beautifully marked with green and yellow, while there are some almost wholly black or brownish olive. But in all varieties, it has one beauty that never seems to change, and that is the eye, which is decorated with a ring of brilliant crimson. It is a small fish, only a few inches in length, and is to be caught in precisely the same manner as the gobies.

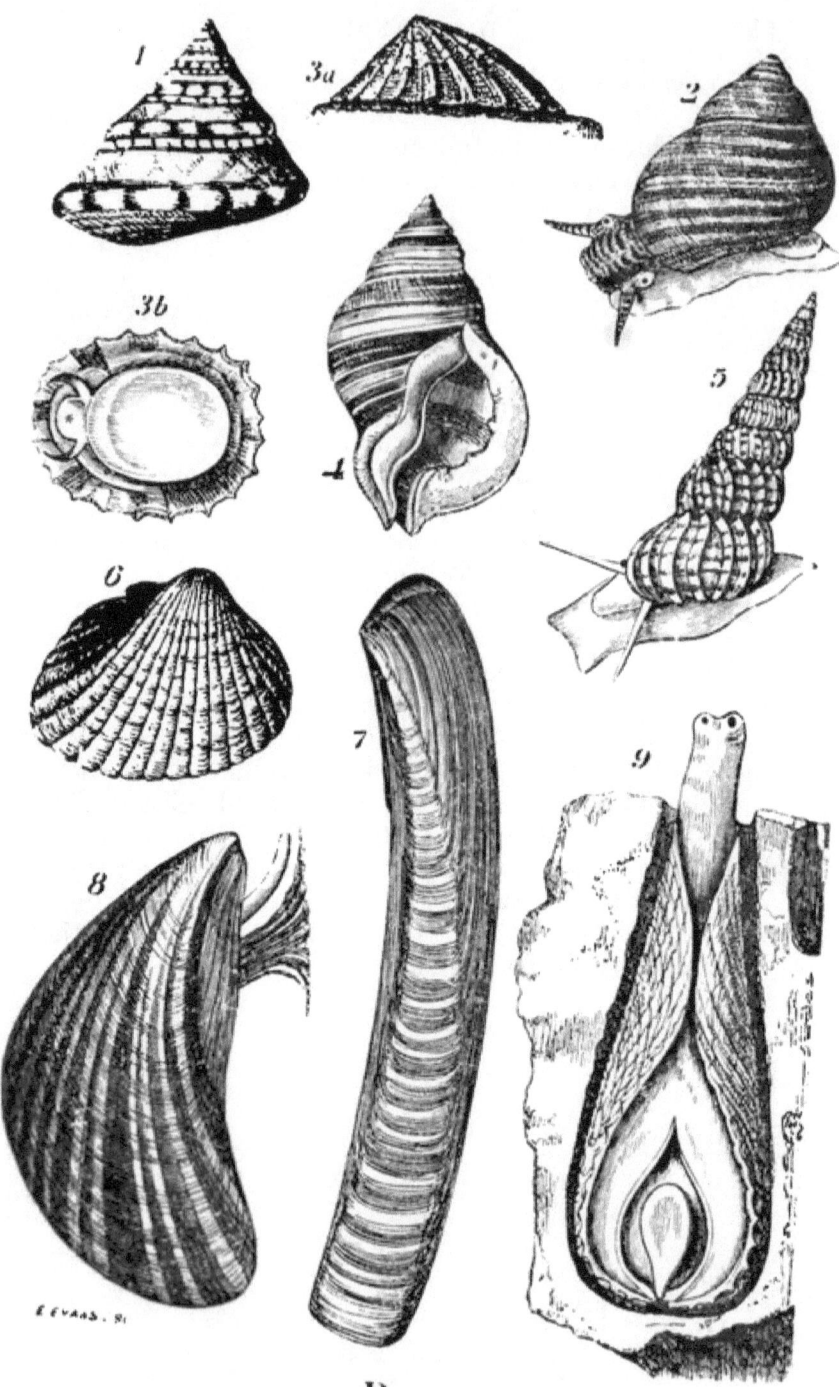

B

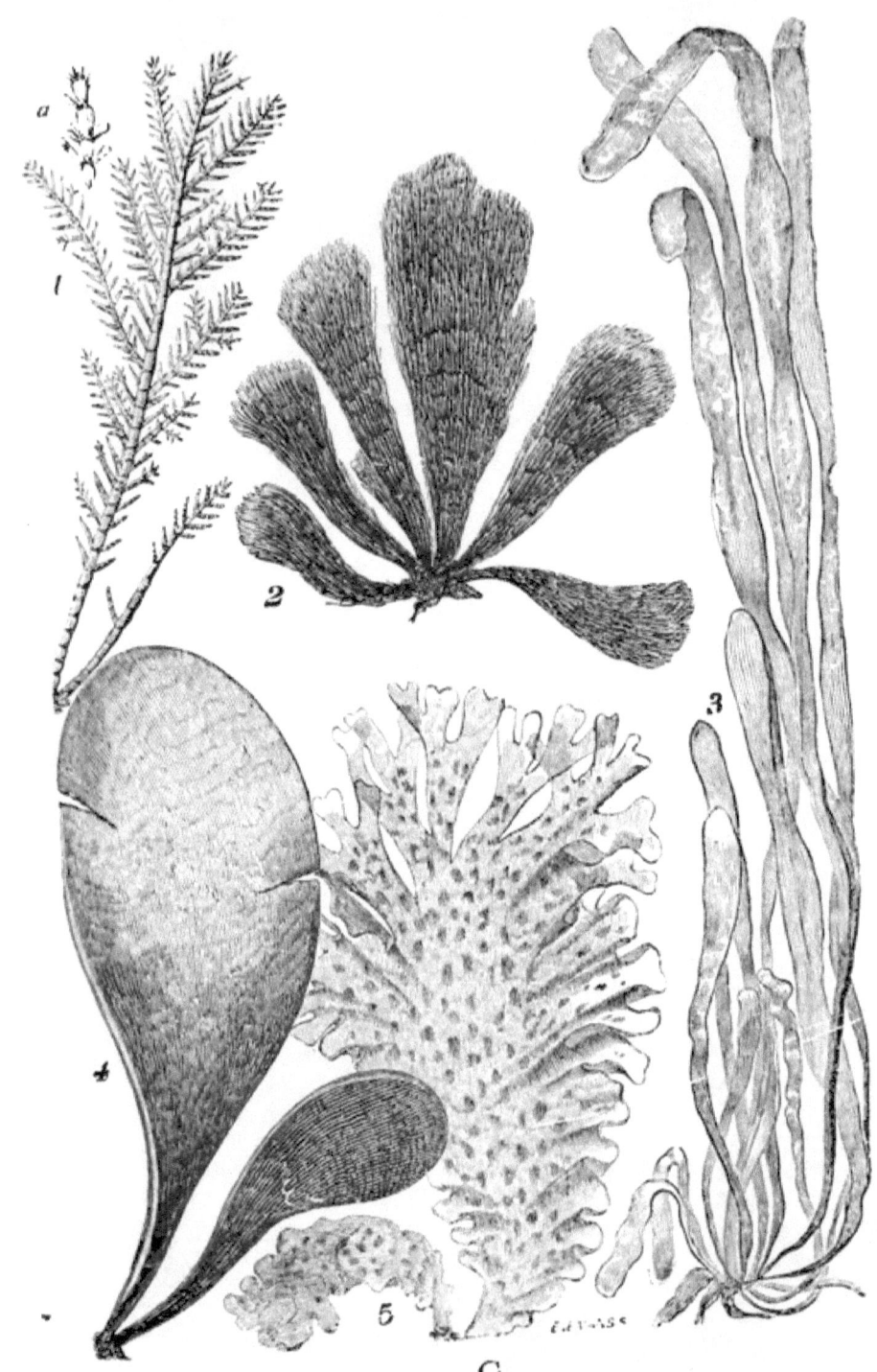

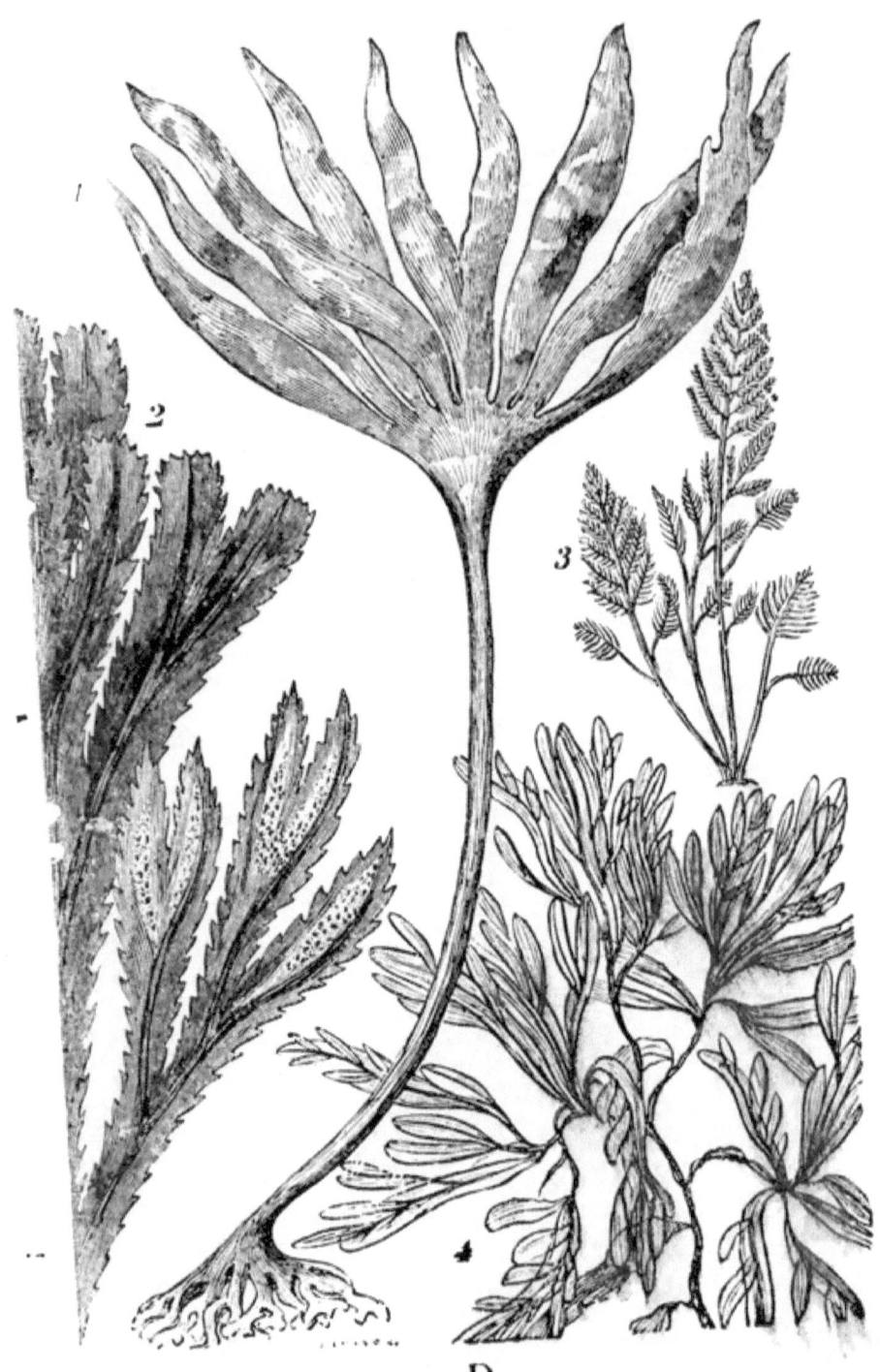

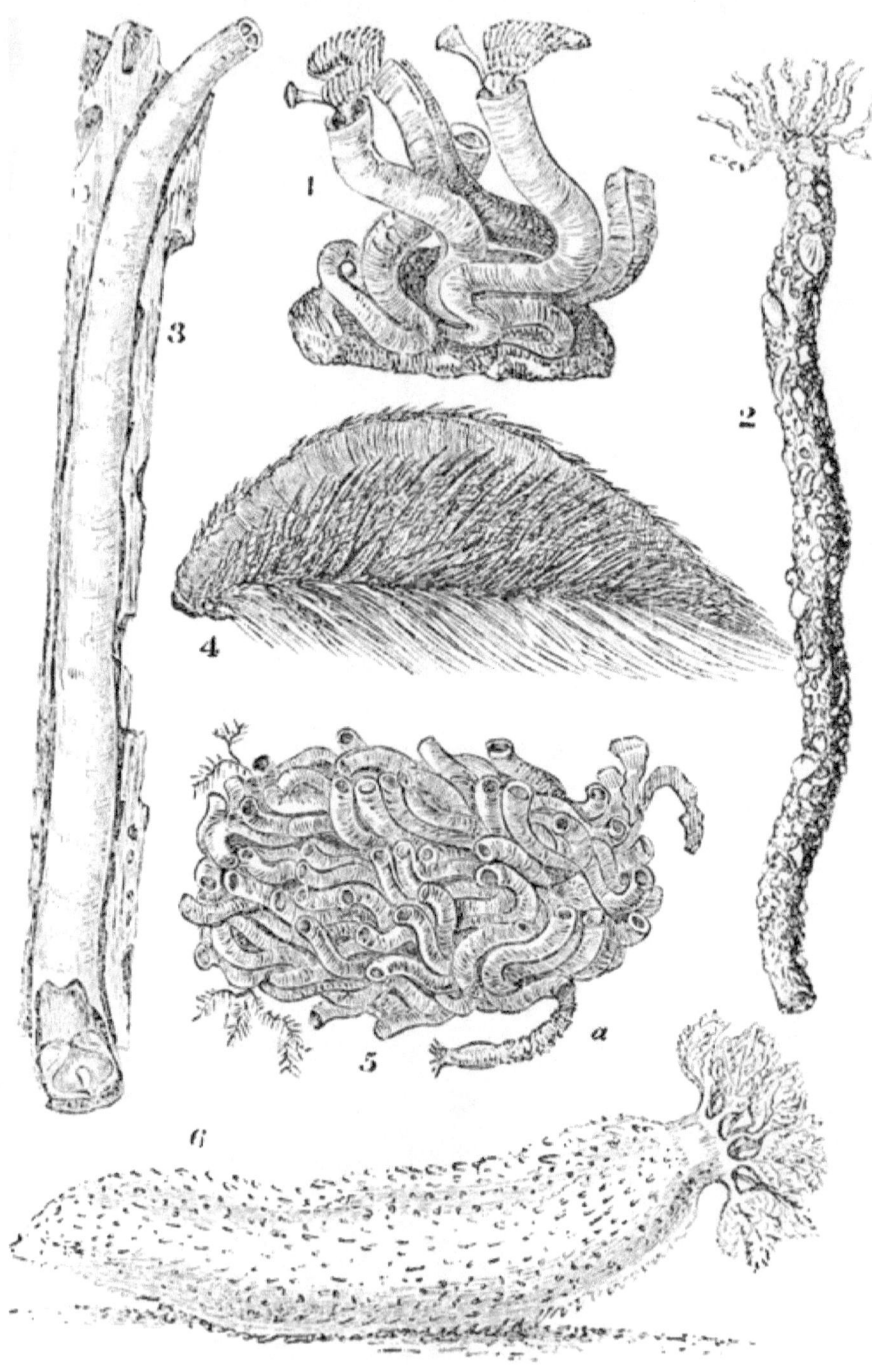

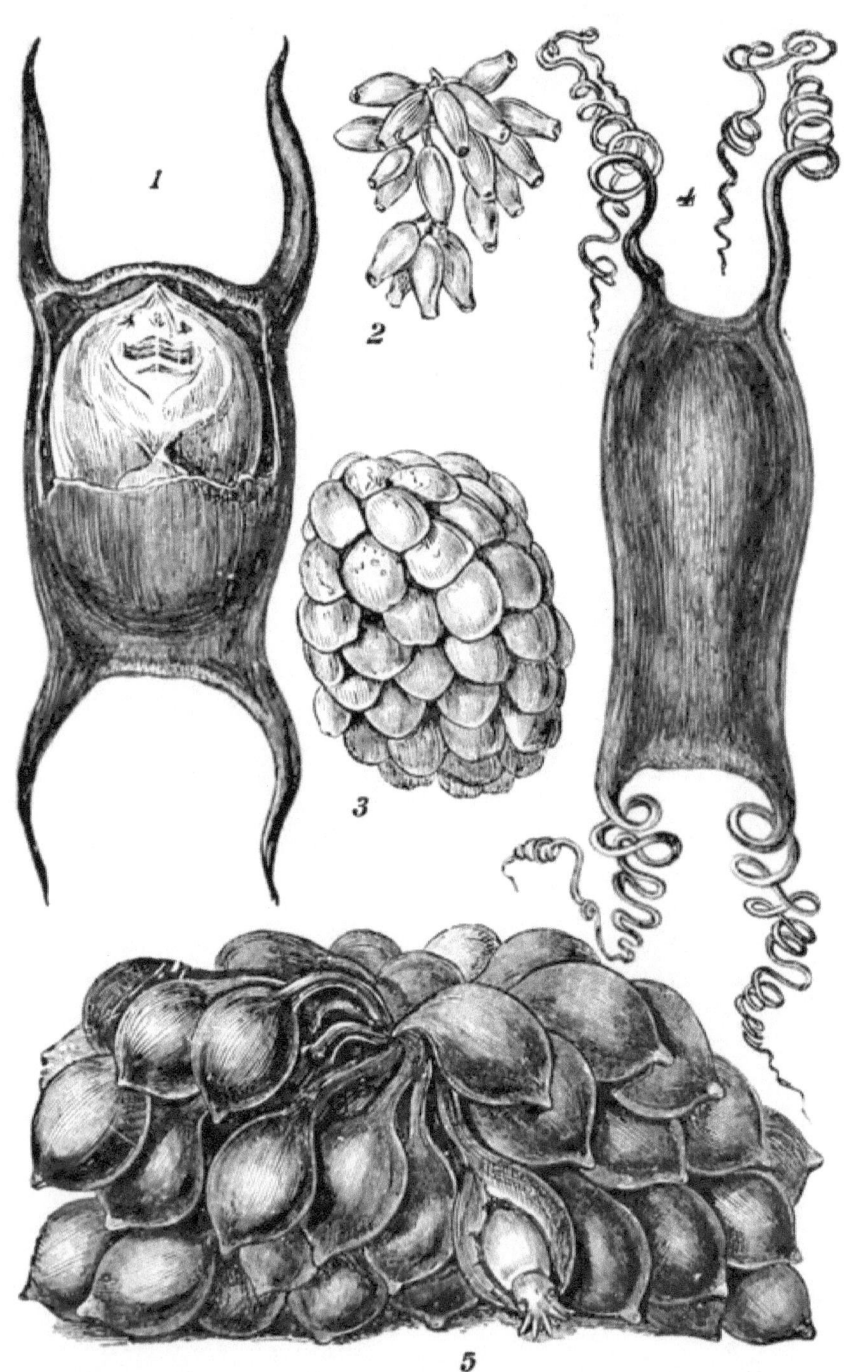

H

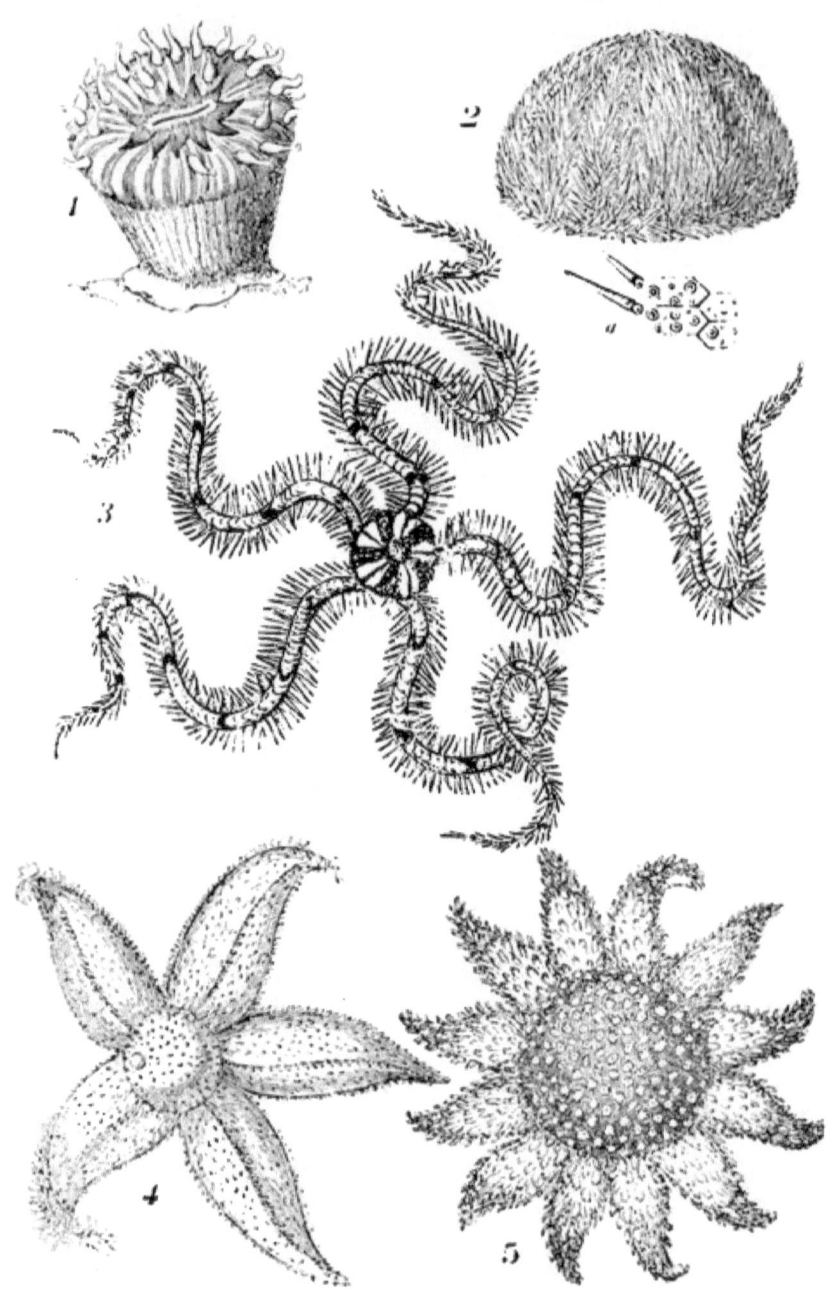

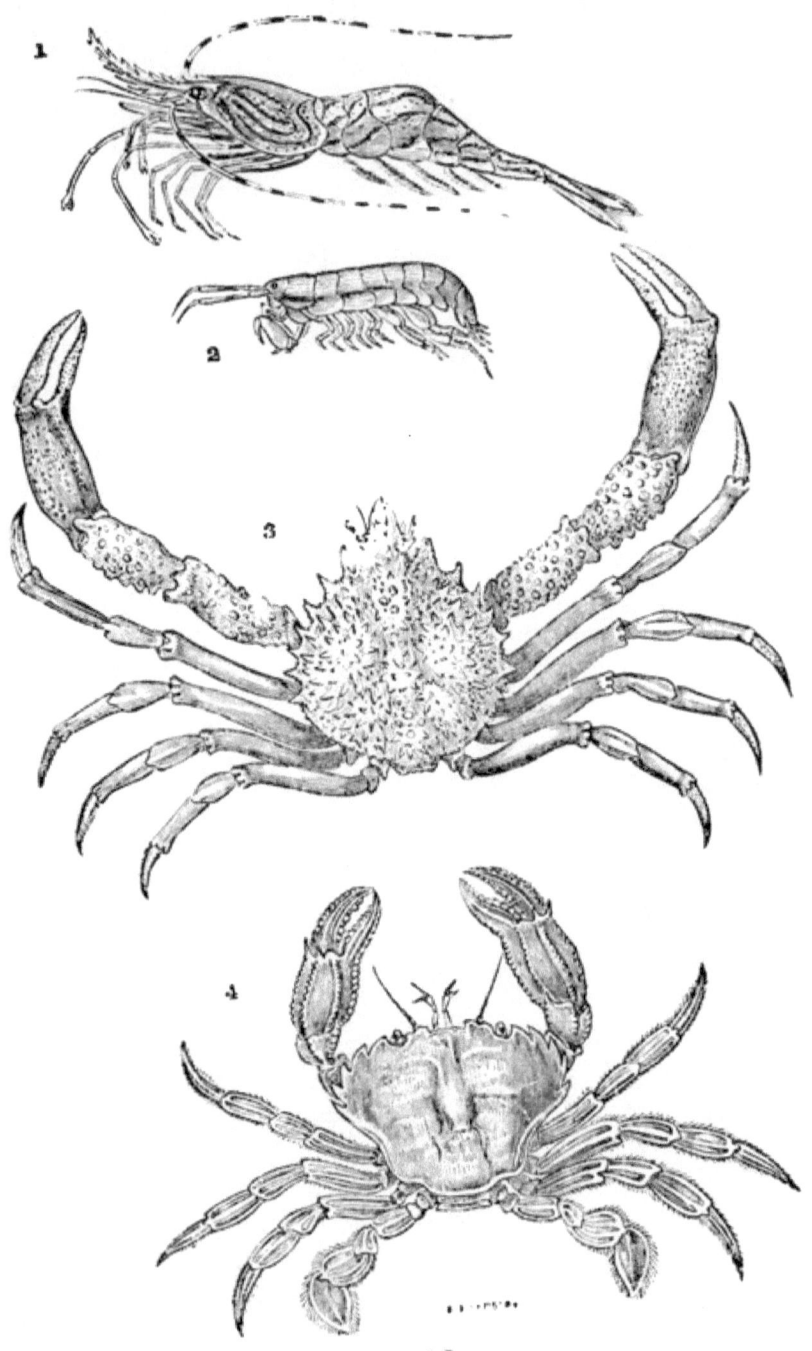

M

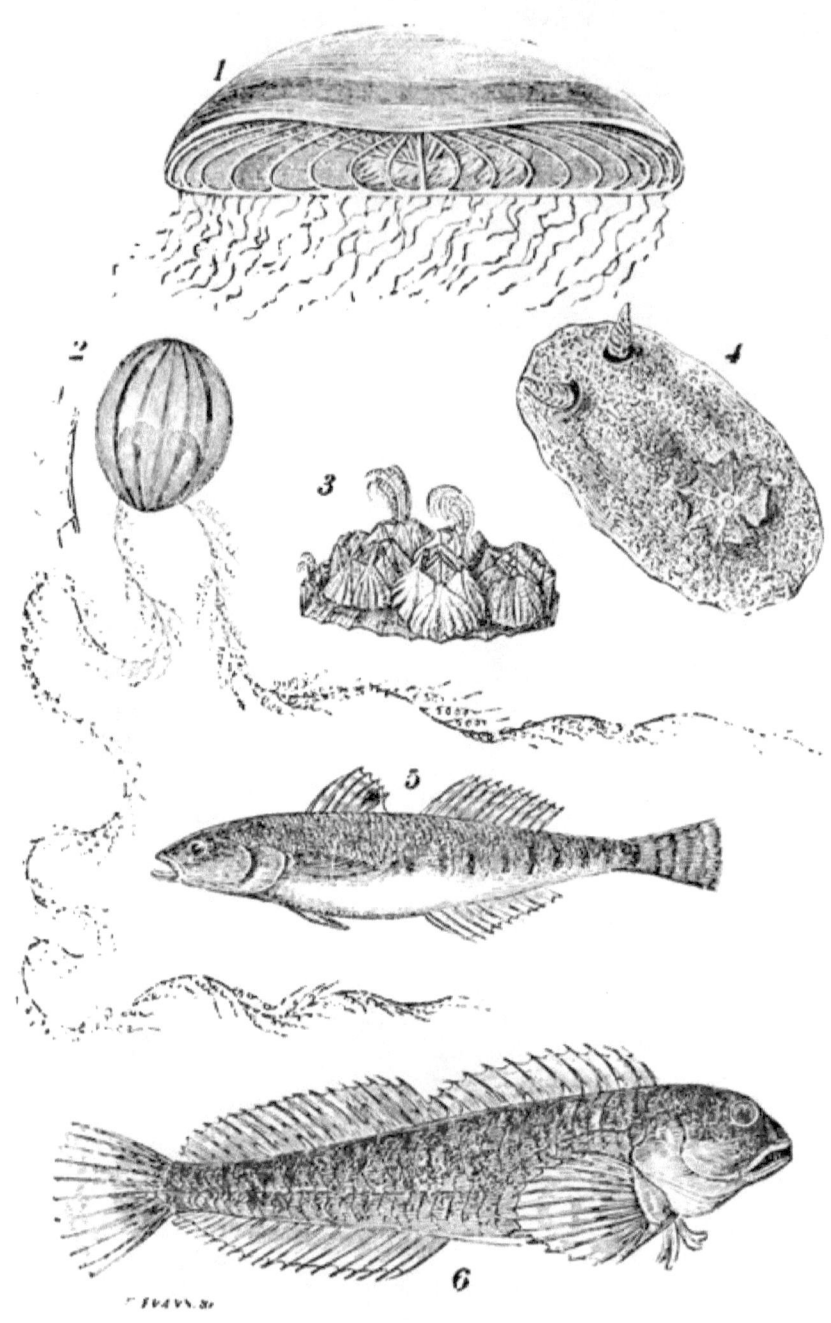

INDEX.

Acorn-shell, 101.
Alcyonium, 75.
Algæ, 27.
Alva, 46.
Anemones, 61.
——— Daisy, 74.
——— smooth, 61.
——— thick-haired, 63.
Annelids, 93.
Aphrodite, 99.
Auks, 7.

Balanus, 101.
Barnacle, 104.
Berniclo-goose, 104.
Beroe, 107.
Birds, 1.
Bird's-head, 78.
Bladder-wrack, 29.
Blenny, 128.
Bryopsis, 42.

Carrageen, 39.
Cephalopods, 53.
Chiton, 24.
Chlorosperms, 41.
Chorda, 32.
Chylocladia, 36.
Cirrhopoda, 101.
Cladophora, 43.
Cockle, 19.
Cod, 57.
Coralline, 37.
Cormorant, 4.

Coryne, 77.
Cowry, 15.
Crab, edible, 112.
——— green, 111.
——— hermit, 112.
——— spider, 114.
——— swimming, 113.
"Crass," 68.
Crustaceans, 111.
Cuttle, 52.

Dead-man's Fingers, 76.
Delesseria, 38, 39.
Dictyota, 33.
Dog-fish, 58.
Doris, 25.
Dunlin, 9.

Echinodermata, 97.
Ectocarpus, 33.
Enteromorpha, 46.

Fish, 126.
Flustra, 74.
Fucus, 29.
Furcellaria, 40.

Gannet, 6.
Goby, 127.
Grapes, Sea, 51.
Griffithsia, 41.
Guillemot, 7.
Gulls, 1.

Holothuria, 91.

Iridæa, 40.
Irish Moss, 40.

Jelly-fish, 106.

Laminaria, 31.
Laver, 44.
Lepralia, 80.
Limpet, 14.
Lobster, 118.
Lug-worm, 94.

Madrepore, 74.
Medusa, 109.
Mermaid's Purse, 58.
Mouse, Sea, 99.
Mussel, 22.

Nautilus, 53.
Nereis, 118.
Nitophyllum, 41.
Noctiluca, 110.
Nudibranchs, 25.

Oar-weed, 31.

Peacock's-tail, 33.
Periwinkle, 18.
Pholas, 20.
Plocamium, 39.
Plumularia, 76.
Polysiphonia, 35.
Porpesse, 10.
Porphyra, 45.
Ptilota, 41.
Prawns, 124.

Puffin, 8.
Purpura, 16.
——— eggs of, 51.
Purre, 10.

Razor-shell, 20.
Rhodosperms, 35.

Sabella, 96.
Sand-skipper, 126.
Scallop, 23.
Serpula, 97.
Sertularia, 76.
Shanny, 128.
Shells, 13.
Shrimps, 122.
Skate, 39.
Star-fishes, 81.
——————— brittle, 87.
——————— five-finger, 81.
——————— sun, 86.

Tansy, 128.
Teredo, 22.
Terebella, 95.
Terns, 3.
Thread-capsules, 70.
Tides, 28.
Tops, 17.

Ulva, 44.
Urchins, 87.

Wentletrap, 19.
Whelk, 15.
——— eggs of, 50.

Zostera, 46.

INDEX TO PLATES.

A (*Front.*)
1. Chylocladia articulata.
2. Ectocarpus siliculosus.
3. Padina pavonia. — (Peacock's-tail.)
4. Porphyra laciniata. — (Purple Laver, or Sloke.)
5. Dictyota dichotoma.

B
1. Trochus ziziphinus. — (Pearly Top.)
2. Littorina littoralis. — (Periwinkle.)
3. Patella vulgaris. — (Limpet.) Ditto, showing under-side.
4. Purpura lapillus.
5. Scalaria communis. — (Common Wentletrap.)
6. Cardium edule. — (Common Cockle.)
7. Solen ensis. — (Razor-shell.)
8. Mytilus edulis. — (Mussel.)
9. Pholas dactylus.

C
1. Corallina officinalis. — (Common Coralline.) A portion of frond, with terminal ceramidium.
2. Cladophora arcta.
3. Enteromorpha compressa. — (Sea Grass.)
4. Iridæa edulis. — (Dulse, or dillosk.)
5. Nitophyllum punctatum.

D
1. Laminaria digitata. — (Tangle.)
2. Fucus serratus. — (Notched wrack.)
3. Bryopsis plumosa.
4. Delesseria hypoglossum.

E
1. Sertularia filicula.
2. Flustra foliacea.
3. Coryne pusilla.
4. Actinia mesembryanthemum.
5. Ditto.
6. Lepralia ciliata.
7. Cellularia avicularis.
8. Alcyonium digitatum.
9. Bunodes crassicornis.

F
1. Serpula contortuplicata.
2. Tube of Terebella.
3. Teredo navalis. — (Ship-worm.)
4. Aphrodite, or Halithea aculeata. — (Sea-mouse.)
5. Sabella.
6. Holothuria. — (Sea-cucumber.)

G
Dunlin. — See p. 9.

H
1. Egg of Skate.
2. Eggs of Purpura.
3. Egg-cluster of Whelk.
4. Egg of Dog-fish.
5. Egg-cluster of Sepia, or Cuttle.

J

1. Fucus nodosus.
2. Zostera marina.—(Grass-wrack or Alva.)
3. Delesseria sanguinea.
4. Furcellaria fastigiata.
5. Chondrus crispus.—(Irish, or Carrageen Moss.)
6. Fucus vesiculosus.—(Bladder-wrack.)

K

1. Griffithsia setacea. A fruit magnified.
2. Polysiphonia urceolata. A fruit magnified.
3. Plocamium coccineum. A portion magnified.
4. Rhodymenia bifida.
5. Ptilota plumosa.
6. Ulva latissima.—(Green Laver, or Sloke.)

L

1. Caryophyllia Smithii.—(Common Madrepore.)
2. Echinus Sphære.—(Urchin, or Sea-egg.)
3. Ophiocoma rosula.—(Brittle-star.)
4. Uraster rubens.—(Five-finger star.)
5. Solaster papposa.—(Sun-star.)

M

1. Pandalus annulicornis.—(Æsop Prawn.)
2. Sandskipper.
3. Maia squinado.—(Spider-crab.)
4. Portunus puber.—(Velvet swimming Crab, or Velvet Fiddler.)

N

1. Æquorea.—(Medusa, or Jelly-fish.)
2. Cydippe pileus.—(Beroe, or Egg Jelly-fish.)
3. Balanus.—(Acorn-shell, or Acorn barnacle.)
4. Doris ptilota.—(Naked-gilled Sea-slug.)
5. Gobius unipunctatus.—(One-spotted Goby.)
6. Blennius pholis.—(Shanny.)

THE END.

www.ingramcontent.com/pod-product-compliance
Lightning Source LLC
Chambersburg PA
CBHW030312170426
43202CB00009B/977